Richard Owen, William John Broderip

Memoir of the Dodo

The bird without wings

Richard Owen, William John Broderip

Memoir of the Dodo
The bird without wings

ISBN/EAN: 9783742840639

Manufactured in Europe, USA, Canada, Australia, Japa

Cover: Foto ©berggeist007 / pixelio.de

Manufactured and distributed by brebook publishing software
(www.brebook.com)

Richard Owen, William John Broderip

Memoir of the Dodo

Whitmore pinxt.   M & N Hanhart imp.   J. Reinholme lith.

MEMOIR

ON

THE DODO

(*Didus ineptus*, Linn.).

BY

RICHARD OWEN, F.R.S.,

WITH AN

HISTORICAL INTRODUCTION

BY THE LATE

WILLIAM JOHN BRODERIP, F.R.S.

LONDON:
PRINTED BY TAYLOR AND FRANCIS, RED LION COURT, FLEET STREET.
1866.

TO

THE HON. ADOLPHUS F. O. LIDDELL, Q.C.

My dear Neighbour,

If our accomplished and lamented friend, Mr. Broderip, had been spared to see the evidences of the extinct bird of the Mauritius described in the following pages, he would probably have taken a more direct share in the present work, and he certainly would have felt equal pleasure with myself in inscribing it to you, in whose society we so often enjoyed pleasant and instructive discourse in the sylvan walks and tranquil shades of Sheen.

Believe me,

Very sincerely yours,

RICHARD OWEN.

Sheen Lodge, Richmond Park,
August 1866.

CONTENTS.

THE DODO

(*Didus ineptus*, LINN.).

§ 1. *Historical Introduction.*

THE DODO has long been one of the "Curiosities of Natural History," through the singularity of its recorded shape, and the paucity of the material evidences of the bird. The head and foot in the Ashmolean Museum at Oxford, and the foot in the British Museum, were all the parts of the bird known to the author of the admirable article "DODO" at the date of its publication in the 'Penny Cyclopædia'[1].

The history of the bird to that date is so conscientiously and exhaustively worked out by my lamented friend, that, instead of paraphrasing or amplifying it, I here give it in Mr. Broderip's own words.

"*Written and Pictorial Evidence.*—In the voyage to the East Indies, in 1598, by Jacob Van Neck and Wybrand van Warwijk (small 4to, Amsterdam, 1648), there is a description of the *Walgh-vogels* in the Island of Cerne, now called Mauritius, as being as large as our swans, with large heads, and a kind of hood thereon; no wings, but, in place of them, three or four black little pens (pennekens), and their tails consisting of four or five curled plumelets (pluymkens) of a greyish colour. The breast is spoken of as very good, but it is stated that the voyagers preferred some Turtle-doves that they found there. The bird appears with a tortoise near it (fig. 1), in a small engraving, one of six which form the prefixed plate.

"In the frontispiece to De Bry (Quinta Pars Indiæ Orientalis, &c., M.DCI.), surmounting the architectural design of the titlepage, will be found, we believe, the earliest engravings of the Dodo. A pair of these birds stand on the cornice on each side, and the following cut (fig. 2) is taken from the figure on the left hand.

Fig. 2.

Tortoise and Walgh-vogel, of the Mauritius (Van Neck and Wybrand, 1598). From plate 2 of Van Neck's Voyage.

Fig. 1.

Dodo
(De Bry, 1601).

[1] By WILLIAM JOHN BRODERIP, Esq., F.R.S. The part containing the article was published in 1836, the volume (IX.) appeared in 1837.

" In De Bry's 'Descriptio Insulæ De Cerne a nobis Mauritius dictæ' is the following account :—'Cærulean Parrots also are there in great numbers, as well as other birds ; besides which there is another larger kind, greater than our swans, with vast heads, and one half covered with a skin, as it were, hooded. These birds are without wings, in the place of which are three or four rather black feathers (quarum loco tres quatuorve pennæ nigriores prodeunt). A few curved delicate ash-coloured feathers constitute the tail. These birds we called *Walck-Vogel*, because the longer they were cooked the more unfit for food they became (quod quo longius seu diutius elixarentur, plus lentescerent et esui ineptiores fierent). Their bellies and breasts were nevertheless of a pleasant flavour (saporis jucundi) and easy of mastication. Another cause for the appellation we gave them was the preferable abundance of Turtle-doves which were of a far sweeter and more grateful flavour.' It will be observed that the bill in De Bry's figure is comparatively small.

" Clusius, in his 'Exotica' (1605), gives a figure, here copied " (note [1], p. 4), " which, he says, he takes from a rough sketch in a journal of a Dutch voyager who had seen the bird in a voyage to the Moluccas in the year 1598.

" The following is Willughby's translation of Clusius, and the section is thus headed : 'The Dodo, called by Clusius *Gallus gallinaceus peregrinus*, by Nieremberg *Cygnus cucullatus*, by Bontius *Dronte*.' 'This exotic bird, found by the Hollanders in the island called Cygnæa or Cerne (that is the Swan Island) by the Portuguese, Mauritius Island by the Low Dutch, of thirty miles' compass, famous especially for black ebony, did equal or exceed a swan in bigness, but was of a far different shape ; for its head was great, covered as it were with a certain membrane resembling a hood : beside, its bill was not flat and broad, but thick and long ; of a yellowish colour next the head, the point being black. The upper chap was hooked ; in the nether had a bluish spot in the middle between the yellow and black part. They reported that it is covered with thin and short feathers, and wants wings, instead whereof it hath only four or five long black feathers ; that the hinder part of the body is very fat and fleshy, wherein for the tail were four or five small curled feathers, twirled up together, of an ash colour. Its legs are thick rather than long, whose upper part, as far as the knee, is covered with black feathers ; the lower part, together with the feet, of a yellowish colour ; its feet divided into four toes, three (and those the longer) standing forward, the fourth and shortest backward : all furnished with black claws. After I had composed and writ down the history of this bird with as much diligence and faithfulness as I could, I happened to see in the house of Peter Pauwius, primary professor of physic in the University of Leyden, a leg thereof cut off at the knee, lately brought over out of Mauritius his island. It was not very long, from the knee to the bending of the foot being but little more than four inches, but of a great thickness, so that it was almost four inches in compass, and covered with thick-set scales, on the upper side broader, and of a yellowish colour, on the under (or back side of the leg) lesser and dusky. The upper side of the toes was also covered with broad scales, the under side wholly callous. The toes were short for so thick a leg: for the length of the greatest or middlemost toe to the nail did not much exceed two inches, that of the other toe next to it scarce came up to two inches: the back toe fell something short of an inch and a half; but the claws of all were thick, hard, black, less than **an inch long; but that of the back toe longer than the rest, exceeding an inch.** The mariners, in their dialect, gave this bird the name *Walgh-Vogel*, that is, a nauseous or yellowish[1] bird; partly because after long boiling its flesh became not tender, but continued hard and of a difficult concoction, excepting the breast and gizzard, which they found **to be** of no bad relish, partly because they could easily get many Turtle-doves, which were much more delicate and pleasant to the palate. Wherefore it was no wonder that in comparison of those they despised this, and said they could be well content without it. **Moreover, they said that they found certain stones** in its gizzard, *and no wonder, for all other birds, as well as these, swallow stones to assist them in grinding their meat.*' Thus far Clusius.

" In the voyage of Jacob Heemskerk and Wolfert Harmanz to the East Indies, in **1601, 1602,** 1603 (small 4to, Amsterdam, 1648), folio 19, the Dod-aarsen (Dodos) are enumerated among **the birds** of the Island of 'Cerne, now Mauritius'; **and** in the 'Journal of the East Indian Voyage of Willem Ysbrantsz Bontekoe van Hoorn, comprising many wonderful and perilous things that happened to him'—from 1618 to 1625 (small 4to, Utrecht, 1649)—**under the head of the 'Island of Mauritius or Maskarinas,'** mention is made (page 6) of the Dod-eersen, which had small wings, but could not fly, and were so fat that they scarcely could go.

" Herbert, in his Travels (1634), gives a figure or rather figures of a bird that he calls ' **Dodo,'** and the following account:—' The Dodo comes first to our description, here, and in Dygarrois (and no where else, that ever I could see or heare of, is generated the **Dodo).** (A Portuguine name it is, and has reference to her simpleness), a bird which for shape and rarenesse might be called a Phœnix (wer't in Arabia); her body is round and extreame fat, her slow pace begets that corpulencie; few of them weigh lesse than fifty pound: better to the eye than the stomack; greasie appetites may perhaps commend them, but to the indifferently curious nourishment, but prove offensive. Let's take her **picture:** her visage darts forth melancholy, as sensible of nature's injurie in framing so **great and massie a body** to be directed by such small and complementall wings, **as are** unable to hoise her from the ground, serving only to prove her a bird; **which otherwise** might be doubted of: her head is variously drest, the one halfe hooded **with downy** blackish feathers; the other perfectly naked; of a whitish hue, as if a transparent lawne had **covered it: her bill** is very howked and bends downwards, the thrill or **breathing place is in the midst of it; from which part to the end, the colour is a light greene mixt with a** pale yellow; her eyes be round and small, and bright as diamonds;

[1] " "So in Willughby, but the print is somewhat indistinct, and there may be error. In the original the words are ' *Walgh-Vogel*, hoc est, nauseam movens avis, partim quod,' &c., the word therefore is an interpolation."

her cloathing is of finest downe, such as you see in goslins; her trayne is (like a China beard) of three or foure short feathers; her legs thick, and black, and strong; her tallons or pounces sharp; her stomack fiery hot, so as stones and iron are easily digested in it; in that and shape, not a little resembling the Africk oestriches: but so much, as for their more certain difference I dare to give thee (with two others) her representation.' (4th ed. 1677.)

"Nieremberg's description (1655) may be considered a copy of that of Clusius, and indeed his whole work is a mere compilation. As we have seen above, he names the bird *Cygnus cucullatus*.

"In Tradescant's catalogue ('Musæum Tradescantianum; or, a Collection of Rarities preserved at South Lambeth, near London, by John Tradescant,' London, 1656, 12mo), we find among the 'Whole Birds'—'Dodar, from the island Mauritius; it is not able to flie being so big.' That this was a Dodo there can be no doubt; for we have the testimony of an eye-witness, whose ornithological competency cannot be doubted, in the affirmative. Willughby at the end of his section on 'The Dodo,' and immediately beneath his translation of Bontius, has the following words: 'We have seen this bird dried, or its skin stuft in Tradescant's cabinet.' We shall, hereafter, trace this specimen to Oxford.

"Jonston (1657) repeats the figure of Clusius, and refers to his description and that of Herbert.

"Bontius, edited by Piso (1658), writes as follows: '*De Dronte* aliis *Dod-aers*.' After stating that among the islands of the East Indies is that which is called Cerne by some, but Mauritius 'a nostratibus,' especially celebrated for its ebony, and that in the said island a bird 'miræ conformationis' called *Dronte* abounds, he proceeds to tell us—we take Willughby's translation—that it is 'for bigness of mean size between an ostrich and a turkey, from which it partly differs in shape, and partly agrees with them, especially with the African ostriches, if you consider the rump, quills, and feathers: so that it was like a pigmy among them, if you regard the shortness of its legs. It hath a great, ill-favoured head, covered with a kind of membrane resembling a hood; great black eyes; a bending, prominent, fat neck; an extraordinary long, strong, bluish-white bill, only the ends of each mandible are of a different colour, that of the upper black, that of the nether yellowish, both sharp-pointed and crooked. It gapes huge wide as being naturally very voracious. Its body is fat, round, covered with soft grey feathers, after the manner of an ostriches: in each side, instead of hard wing-feathers or quills, it is furnished with small, soft-feathered wings, of a yellowish ash-colour; and behind, the rump, instead of a tail, is adorned with five small curled feathers of the same colour. It hath yellow legs, thick, but very short; four toes in each foot, solid, long, as it were

---

[1] These and other grotesque figures, which may be seen, copied, in Strickland's History of the Dodo ('Dodo and its Kindred,' 4to, 1848), from the old authors cited by Broderip, are mere matters of curiosity, and are here omitted as devoid of scientific value.

scaly, armed with strong, black claws. It is a slow-paced and stupid bird, and which easily becomes a prey to the fowlers. The flesh, especially of the breast, is fat, esculent, and so copious, that three or four Dodos will sometimes suffice to fill an hundred seamens' bellies. If they be old, or not well boiled, they are of difficult concoction, and are salted and stored up for provision of victual. There are found in their stomachs stones of an ash colour, of divers figures and magnitudes; yet not bred there, as the common people and seamen fancy, but swallowed by the bird; as though by this mark also nature would manifest that these fowl are of the ostrich kind, in that they swallow any hard things, though they do not digest them.'

" It appears from Adam Olearius (Die Gottorfische Kunst Kammer, 1666), that there was a head to be seen in the Gottorf Museum; but the figure (tab. 13. f. 5) is very like that of Clusius. It is mentioned as the head of the *Walch-Vogel*, and Clusius is referred to. In the plate the head is shaded, and has a more finished appearance: the rest of the bird is in outline[1].

" Grew ('Musæum Regalis Societatis; or a catalogue and description of the natural and artificial rarities belonging to the Royal Society,' London, folio, 1681), at p. 68, thus describes the bird which is the subject of our inquiry. 'The leg of a Dodo; called *Cygnus cucullatus* by Nierembergius; by Clusius, *Gallus gallinaceus peregrinus*; by Bontius called *Dronte*, who saith that by some it is called (in Dutch) *Dod-aers*, largely described in Mr. Willughby's Ornithol. out of Clusius and others. He is more especially distinguished from other birds by the membranous hood on his head, the greatness and strength of his bill, the littleness of his wings, his bunchy tail, and the shortness of his legs. Abating his head and legs, he seems to be much like an ostrich, to which also he comes near as to the bigness of his body. He breeds in Mauris's Island. The leg here preserved is covered with a reddish-yellow scale. Not much above four inches long, yet above five in thickness, or round above the joints, wherein, though it be inferior to that of an Ostrich or Cassowary, yet, joined with its shortness, may render it of almost equal strength.' At p. 73, there is the following notice :—'The head of the Man of War, called also Albitrosse; supposed by some to be the head of a Dodo, but it seems doubtful. That there is a bird called the Man of War is commonly known to our seamen; and several of them who have seen the head here preserved, do affirm it to be the head of that bird, which they describe to be a very great one, the wings whereof are eight feet over. And Ligon (Hist. of Barbad. p. 61), speaking of him, saith, that he will commonly fly out to sea to see what ships are coming to land, and so return. Whereas the Dodo is hardly a volatile bird, having little or no wings, except such as those of the Cassowary and the Ostrich. Besides, although the upper beak of this bill doth much resemble that of the Dodo, yet the nether is of a quite different shape; so that this either is not the head of a Dodo, or else we have nowhere a true figure of it.' Grew then gives a very lengthened description of the skull which is figured by him

[1] This head, in the condition of a skull, has subsequently been discovered at Copenhagen.—R. O.

6

(tab. 6), and intituled ' Head of the Albitros,' as it doubtless was. The leg above mentioned is that now preserved in the British Museum, where it was deposited with the other specimens described by Grew, when the Royal Society gave their ' rarities' to that national establishment. Grew was a well qualified observer, and much of this description implies observation and comparison; indeed, though he does not refer to it, there is no reason for supposing that Grew was not familiar with Tradescant's specimen.

"Charleton also (Onomasticon, 1668) speaks of the Dodo Lusitanorum (*Cygnus cucullatus*, Willughby and Ray), and asserts that the Museum of the Royal Society of London contained a leg of the Dodo. This was evidently the leg above alluded to.

"We now proceed to trace the specimen which was in the Museum Tradescantianum. There were, it seems, three Tradescants, grandfather, father, and son. The two former are said to have been gardeners to Queen Elizabeth, and the latter to Charles I. There are two portraits to the ' Museum,' one of ' Joannes Tradescantus pater,' and the other of ' Joannes Tradescantus filius,' by Hollar. These two appear to have been the collectors: for John Tradescant, the son, writes in his address, ' to the ingenious reader' that ' he was resolved to take a catalogue of those varieties and curiosities which my father had sedulously collected and my selfe with continued diligence have augmented, and hitherto preserved together.' This John Tradescant, the son, must have been the Tradescant with whom Elias Ashmole boarded for a summer when Ashmole agreed to purchase the collection, which was said to have been conveyed to Ashmole by deed of gift from Tradescant and his wife. Tradescant died soon after, and Ashmole, in 1662, filed a bill in Chancery for a delivery of the curiosities. The cause is stated to have come to a hearing in 1664; and, in 1674, Mrs. Tradescant delivered up the collection pursuant to a decree in Chancery, and afterwards (April, 1678, some say) was found drowned in her own pond. Ashmole added to the collection, and presented it to the University of Oxford, where it became the foundation of the Ashmolean Museum. That the entire ' Dodo' went to Oxford with the rest of Tradescant's curiosities there can be no doubt. Hyde (Religionis Veterum Persarum, &c., Historia, 1700) makes particular mention of it as existing in the Museum at Oxford. There, according to Mr. Duncan, it was destroyed in 1755 by order of the visitors, and he thus gives the evidence of its destruction :—

" ' In the Ashmolean Catalogue, made by Ed. Llhwyd, Musæi Procustos, 1684 (Plott being the keeper), the entry of the bird is, " No. 29. *Gallus gallinaceus peregrinus*, Clusii,' &c. In a Catalogue made subsequently to 1755, it is stated "That the numbers from 5 to 46, being decayed, were ordered to be removed at a meeting of the majority of the visitors, Jan. 8, 1755." Among these of course was included the Dodo, its number being 29. This is further shown by a new Catalogue, completed in 1756, in which the order of the visitors is recorded as follows: " Ilis quibus nullus in margine assignatur numerus a Musæo subducta sunt cimelia, annuentibus Vice-Cancellario aliisque Curatoribus ad ea lustranda convocatis, die Januarii 8vo, A.D. 1756." The

7

Dodo is one of those which are here without the number.' (Duncan, "On the Dodo," Zool. Journ. vol. iii. p. 559.)

"We now come to the celebrated painting in the British Museum, a copy of which, by the kind assistance of the officers of the zoological department, who have given us every assistance in prosecuting this inquiry, and who had it taken down for the purpose, we present to our readers[1].

"It has been stated that the painting came into the possession of Sir Hans Sloane, president of the Royal Society, and that it was bought at his sale by Edwards, who, after publishing a plate from it in his Gleanings, presented it to the Royal Society, whence it passed, as well as the foot, into the British Museum. But Mr. Gray informs us that the foot only came with the museum of the Royal Society described by Grew; and that the picture was an especial gift from Edwards. Edwards's copy seems to have been made in 1760, and he himself says—'The original picture was drawn in Holland from the living bird brought from St. Maurice's Island, in the East Indies, in the early times of the discovery of the Indies by the way of the Cape of Good Hope. It was the property of the late Sir Hans Sloane to the time of his death; and afterwards becoming my property I deposited it in the British Museum as a great curiosity. The above history of the picture I had from Sir Hans Sloane and the late Dr. Mortimer, secretary to the Royal Society.'

"M. Morel (Ecrivain Principal des Hôpitaux au Port-Louis de l'Isle de France) writes as follows in his paper 'Sur les oiseaux monstrueux nommés Dronte, Dodo, Cygne Capuchonné, Solitaire, et Oiseau de Nazare, et sur la petite Isle de Sable à 50 lieues environ de Madagascar.' 'These birds, so well described in the second volume of the 'History of Birds,' by M. le Comte de Buffon, and of which M. de Borame has also spoken in his 'Dictionary of Natural History,' under the names of Dronte, Dodo, Hooded Swan (Cygne Capuchonné), Solitary or Wild Turkey (Dinde sauvage) of Madagascar, have never been seen in the isles of France, Bourbon, Rodriguez, or even the Seychelles lately discovered, during more than sixty years since when these places have been inhabited and visited by French colonists. The oldest inhabitants assure every one that these monstrous birds have been always unknown to them.' After some remarks that the Portuguese and Dutch who first overran these islands may have seen some very large birds, such as Emeus or Cassowaries, &c., and described them each after his own manner of observing, M. Morel thus proceeds: 'However this may be, it is certain that for nearly an age (depuis près un siècle) no one has here seen an animal of this species. But it is very probable that before the islands were inhabited, people might have been able to find some species of very large birds, heavy and incapable of flight, and that the first mariners who sojourned there soon destroyed them from the facility with which they were caught. This was what made the Dutch sailors call the bird 'Oiseau de dégoût' (Walck-Voegel), because they were surfeited with the flesh of

---

[1] The outline of the Dodo in this painting is given, of the natural size, in Pl. III. of the present work; the reduced woodcut (tom. cit. p. 51, copied by Strickland, op. cit. p. 28) is, therefore, not here reproduced.—R. O.

it. . . . But among all the species of birds which are found on this isle of sand and on all the other islets and rocks which are in the neighbourhood of the Isle of France, modern navigators have never found anything approaching to the birds above named, and which may be referred to the number of species which may have existed, but which have been destroyed by the too great facility with which they are taken, and which are no longer found excepting upon islands or coasts entirely uninhabited. At Madagascar, where there are many species of birds unknown in these islands, none have been met with resembling the description above alluded to.' (Observations sur la Physique pour l'an 1778, tom. xii. p. 154, notes.)

" Mr. Duncan thus concludes his paper above alluded to :—' Having applied, through the medium of a friend, to C. Telfair, Esq., of Port Louis, in the Mauritius, a naturalist of great research, for any information he could furnish or procure relating to the former existence of the Dodo in that island, I obtained only the following partly negative statement :—

" 'That there is a very general impression among the inhabitants that the Dodo did exist at Rodriguez, as well as in the Mauritius itself; but that the oldest inhabitants have never seen it, nor has the bird or any part of it been preserved in any museum or collection formed in those islands, although some distinguished amateurs in natural history have passed their lives on them, and formed extensive collections. And with regard to the supposed existence of the Dodo in Madagascar, although Mr. Telfair had not received, at the time of his writing to Europe, a reply to a letter on the subject which he had addressed to a gentleman resident on that island, yet he stated that he had not any great expectations from that quarter; as the Dodo was not mentioned in any of his voluminous manuscripts respecting that island, which contained the travels of persons who had traversed Madagascar in all directions, many of them having no other object in view than that of extending the bounds of natural history.'

" We close this part of the case with the evidence of one evidently well qualified to judge, and whose veracity there is no reason to doubt. If this evidence be, as we believe it to be, unimpeachable, it is clear not only that the Dodo existed, but that it was publicly exhibited in London. The lacunæ in the print represent the spaces occasioned by a hole burnt in the manuscript.

" In the 'Sloane MSS.' (No. 1839, 5, p. 108, Brit. Mus.) is the following interesting account by L'Estrange in his observations on Sir Thomas Browne's 'Vulgar Errors.' It is worthy of note that the paragraph immediately follows one on the 'Estridge' (Ostrich).

" 'About 1638, as I walked London streets I saw the picture   of a strange fowl hong out upon a cloth   vas and myselfe with one or two more Gen. in company went in to see it. It was kept in a chamber, and was a great fowle somewhat bigger than the largest Turkey Cock and so legged and footed but stouter and thicker and of a more erect shape, coloured before like the breast of a yong Cock Fesan (pheasant), and on the back of dunn or deare coulour. The keeper called it a Dodo and in the ende of a chimney in the chamber there lay an heap of large pebble stones whereof hee gave it many in our sight, some as big as nutmegs, and the keeper told us shee eats them (conducing to digestion) and though I remember not how farre the keeper was questioned therein yet I am confident that afterwards she cast them all agayne[1].'

" *Evidence arising from Remains.*—The only existing recent remains attributed to the Dodo are, a leg (fig. 4) in the British Museum, and a head (fig. 3) (a cast of which is in the British Museum), and a leg in the Ashmolean Museum at Oxford, the relics most probably of Tradescant's bird. Whether the leg formerly in the museum of Pauw be that at present in the British Museum may be, perhaps, doubtful, though we think with Mr. Gray that they are probably identical; but that the specimen in the British Museum did not belong to Tradescant's specimen is clear, for it existed in the collection belonging to the Royal Society when Tradescant's ' Dodar' was complete.

In the ' Annales des Sciences' (tome xxi. p. 103, Sept. 1830) will be found an account of an assemblage of fossil bones, then recently discovered, under a bed of lava, in the Isle of France, and sent to the Paris Museum. They almost all belonged to a large living species of land-tortoise, called *Testudo indica,* but amongst them were the head, sternum, and humerus of the Dodo. 'M. Cuvier,' adds Mr. Lyell in his ' Principles of Geology,' ' showed me these valuable remains at Paris, and assured me that they left no doubt in his mind that the huge bird was one of the gallinaceous tribe[2].'"

Fig. 3.

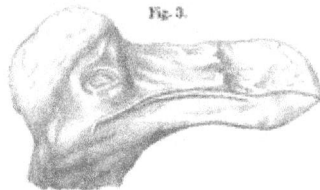

Head of Dodo (specimen in the Oxford Museum), one-third nat. size.

Fig. 4.

Foot of Dodo (specimen in the British Museum), one-third nat. size.

" [1] This curious statement is extracted in the recent edition of Sir Thomas Browne's works by Wilkins: published by Pickering." (8vo, 1836, vol. i. p. 369, vol. ii. 173. The reference, in Strickland (*op. cit.* p. 22), to vol. i. p. 369, is to a Letter by Sir Hamon L'Estrange to Dr. Browne, not containing any allusion to the Dodo.—K. O.] [2] Art. Dodo, Penny Cyclopædia, vol. ix. p. 52 (1837).

c

10

The bones in question were obtained from a cavern in the Island of Rodriguez (Desjardins, Analyse des Travaux de la Soc. d'Hist. Nat. de l'Ile Maurice, 2e année), and belong to the Solitaire (*Pezophaps*), a large extinct brevipennate bird, allied to the Dodo. The other evidences from remains, cited by Broderip, also relate to the Solitaire. Such was the history of the Dodo in 1837.

In the following year I visited Holland, chiefly with a view to ascertain whether there might possibly be any remnant of the bird preserved in the Natural History Museums of that country, and to collect for my friend whatever other evidence, material, written or pictorial, might have escaped his assiduous researches.

My visits to the museums at Leyden, Amsterdam, Utrecht, and the Hague, during which I received every requisite aid from the accomplished Professors and Curators, were productive of only negative results. The little other information I was able to obtain was communicated to Mr. Broderip, who incorporated it in the following "Supplement to his History."

"*Additional evidence relative to the Dodo. By* W. J. Broderip, *Esq., F.R.S.*

"The interest which attaches to any communication relative to an extinct, and, at one time, a doubted species, must be my apology for offering the following addition to the evidences of the existence and habits of the Dodo.

"My old and valued friend Professor Owen presented me, on his return from Holland some time since, with a short thick volume, bearing on its titlepage (not without black letter) the following promise:—

"'C. Plinii Secundi Des wijdt-vermaerden Natuurkondigers vijf Boecken.

Handelen van de Natuur.

I. Van de Menschen.

II. Van de viervoetige en Kruypende Dieren.

III. Van de Vogelen.

IV. Van de Kleyne Beestjes of Ongedierten.

V. Van de Visschen, Oesters, Kreeften, &c.

"'Hier zijn by-gevoeght de Schriften van verscheyden andere oude Authueren de Natuur der Dieren aengaende. En nu in desen laetsten Druck wel het vierde part vermeerdert, uyt verscheyden nieuwe Schrijvers en eygen oudervindinge: en met veel Kopere Platen verziert t'Amsterdam. By Abraham Wolfganon, 1662.'

"The frontispiece presents the artist's notion of the Garden of Eden, with a very Dutch Adam and Eve, the latter with the apple in her hand, while the serpent twined round the tree looks sly and satisfied. Our first parents are surrounded by beasts, and in the foreground is represented a piece of water with waterfowl and ill-shaped fishes.'

"The superscription is 'C. Plinius S. de Menschen, Beesten, Vogelen en Visschen.'

"Mr. Strickland, in his elaborate work on 'The Dodo and its Kindred[1],' in which

[1] London, 4to, Reeve and Co., 1848."

he has done me the honour to adopt the arrangement and the information collected in my article 'Dodo,' in the ' Penny Cyclopædia',' gives some addenda in his postscript to Part I. of his and Dr. Melville's book. ' The first of these,' writes Mr. Strickland, ' is a rare edition of Bontekoe's Voyage, kindly communicated to me by Dr. Bandinel, the **Bodleian** Librarian, entitled "Journael van de acht-jarige aventuerlijcke Reyse **van Willem** Ysbrantsz Bontekoe van Hoorn, gedaen nae Oost-Indien," published in quarto at Amsterdam, by Gillis Joosten Zaagman. There is no date; but from a narrative introduced at the end, it must be subsequent (probably by a year or two) to 1646. The narrative is nearly a verbatim version of the other Dutch editions of Bontekoe; and the only variation of text which concerns us, is in the statement that the underside of the Dodo dragged along the ground, which is here qualified thus:—"sleepte haer de neers *by na* (i. e. *almost*) langs de Aerde." But what gives a peculiar interest to this volume is, that it contains (alone of all the editions of Bontekoe which I have seen) a figure of the Dodo, which I here present.' Then follows the cut.

" 'This highly ludicrous representation,' continues Mr. Strickland, ' is more like a fighting cock than a Dodo; and the black letter of the Dutch text omits to tell us whether this design was due to the pencil of Bontekoe or his publisher Zaagman, or whether it was copied from some contemporary painting now forgotten. But there can be no doubt that this figure **refers** to the true Dodo of Mauritius, and not to the "Solitaire" of Bourbon, with which Bontekoe confounded it.

" 'We may regret that the rudeness of the original woodcut leaves us in the dark as to the nature of the object on which the Dodo appears about to feed. This figure would pass equally well for a testaceous mollusk, or for an arboreal fruit; so that the problem of the Dodo's food seems as far from a solution as ever.'

" In Wolfgangh's publication, p. 480, is the following description:—

" 'Op't Eylandt Mauritius in Oost-Indien, als mede op sommige andere plaetsen gelijck mede in West-Indien, vindt men voegels soo groot als Swanen, die men Dod-aersen of Draaten noemt, sy hebben groote hoofden, en daer op een vlieken in manier van een Kapken, sy hebben geen vleugels, dan in plaetsvan dien, 3 of 4 swarte pennekens, en daer haer staert behoorde te staen, daer Zijn 4 of 5 gekrulde Plúymkens, van graeuwachtige verwe. Sy hebben een dicke ronde Naers, daer uyt het schijnt, dat haer de naem van Dodkers toe gekomen is; in de maergh hebben sy gemeenlijck een Steen van een vuyst groot, dese is bruyn, graeuw van verwe, en vol gaetkens, en hollingheydt, doch soo hart als graeuwe Bentemeer-steen. Het Bootsvolck van *Jacob von Neck*, noemdense Walgh-vogels, om datse die niet recht gaer of murrruw konden koken: of om datse soo veel Tortel-duyven konden bekomen, die leckerder smaeckten, datse van dese Dod-aersen de walgh kregen. Aen 3 of 4 van dese Vogels had al't Scheeps volck van een Schip, voor een maeltijdt genoegh t' eeten: Dese Dod-aersen hebbense oock ingesouten en op de reys mede genomen.'

" ' Vol. ix. p. 47 (1897)."

c 2

"This description may be thus rendered:—

"'In the Island of Mauritius in the East Indies, as also in sundry other places, likewise in the West Indies, men find birds as big as swans, which they call *Dod-aersen* or *Dromtes*. They have large heads, upon the top of which is a skin (a little skin-membrane) in the shape of a cap (little cap). They have no wings, but in the place of them there are three or four black feathers; and there where the tail should be, there are instead four or five curling plumes of a greyish colour. They have a thick round rump, and from this it appears they got the name of Dodaerses. In their stomachs they have commonly a stone as big as a fist; this stone is of a brown-grey colour, and full of little holes and hollows, but as hard as the grey Bentemer stone. The boat's crew of *Jacob van Neck* called them Walgh-vogels (surfeit birds), because they could not cook them till they were done, or make them tender; or because they were able to get so many turtledoves which had a much more pleasant flavour, so that they took a disgust to these birds. Likewise it is said that three or four of these birds are enough to afford a whole ship's company one full meal. Indeed they salted down some of them, and carried them with them on the voyage.'

"At the top of the page in which this passage commences is printed '*Van de Dod-aersen.*' And immediately below it and above the description is a copper-plate of the bird, superscribed '*Dod-aers,*' in engraved italics.

"The engraving of the bird is identical in position and accessories with the woodcut given by Mr. Strickland; but the sharpness of the work and the nature of the plate make the whole much clearer. The object at which the Dodo is looking, as if about to feed, is manifestly a testaceous mollusk with a turbinated shell, and between that and the raised foot of the bird is a half-buried spiny *Echinus.*

"The locality on which the Dodo is walking has the appearance of a strand which the tide has left dry.

"Wolfgangh's account confirms the opinion which I hazarded in the article 'Dodo' in the 'Penny Cyclopædia.'

"'As to the stories of the disgusting quality of the flesh of the bird found and eaten by the Dutch, they will weigh but little in the scale when we take the expression to be, what it really was, indicative of a comparative preference for the turtle-doves there found, after feeding on Dodos *usque ad nauseam.* "Always partridges" has become proverbial, and we find from Lawson how a repetition of the most delicious food palls. "We cooked our supper," says that traveller, "but having neither bread nor salt, our fat turkeys began to be loathsome to us; although we were never wanting of a good appetite, yet a continuance of one diet made us weary," and again, "By the way our guide killed more turkeys, and two polecats, which he eat, esteeming them before fat turkeys."'

"It does not follow that because the Dodo is represented as looking at the *frutti di mari,* he is about to devour them. But if it be granted he is, the admission would not militate against the opinion of those who would place the Dodo between the Struthious and Gallinaceous birds. It is well known that the turkeys in America come down to the shore and feed upon the 'fiddler' crabs; and there would be nothing extraordinary in a quisquilious feeder, such as the Dodo probably was, varying its fruit and vegetable diet occasionally by resorting to such animal substances as it might find on the strand. Common poultry eagerly pick up insects and slugs in the fields, and, in the neighbourhood of tidal rivers and estuaries, may be seen availing themselves of the smaller *mollusca* and *crustacea* left by the retreating tide.

" In my article 'Struthionidæ' under the section 'Didus,' is inserted the following extract from a letter written to me by Professor Owen :—

" ' Whilst at the Hague in the summer of 1848, I was much struck with the minuteness and accuracy with which the exotic species of animals had been painted by Savery and Breughel, in such subjects as *Paradise, Orpheus charming the beasts*, &c., in which scope was allowed for grouping together a great variety of animals. Understanding that the celebrated menagerie of Prince Maurice had afforded the living models to those artists, I sat down one day before Savery's *Orpheus and the beasts*, to make a list of the species, which the picture evinced that the artist had had the opportunity to study alive. Judge of my surprise and pleasure in detecting in a dark corner of the picture (which is badly hung between two windows), the Dodo beautifully finished, showing for example, though but three inches long, the auricular circle of feathers, the scutation of the tarsi, and the loose structure of the caudal plumes. In the number and proportions of the toes and in general form, it accords with Edwards's oil-painting in the British Museum; and I conclude that the miniature must have been copied from the study of a living bird, which, it is most probable, formed part of the Mauritian menagerie.'

" I little thought, when, with his permission, I published this graphic product of my kind friend's pen, what was in store for me. Not long afterwards, a friend informed me that he had seen a picture at a dealer's painted by one of the Saverys, and that he was pretty sure there was a Dodo in one corner of it. I sent for the picture, and there, sure enough, in the right-hand corner, and consequently to the left of the spectator, was the bird, in all the beauty of its ugliness. The Dodo stands on one foot with its back to the spectator, and turning round its head, which is represented with the huge bill picking the other uplifted foot. Like all the rest of the birds in this picture, which bears the name of Roland Savery, the Dodo is highly finished. The picture is now in my possession[2]."

The figure 2 in Plate I. is a faithful copy of the bird as represented in it.

Whilst on a visit to Sion House I was unexpectedly gratified by finding, in a small oil-painting in the long gallery, an unequivocal and original representation of the Dodo, in an attitude different from that of any of the figures of the living bird by Roland Savery, and evidently by another master. I lost no time in communicating

" [1] Penny Cyclopædia, vol. xxiii. (1842)."
" [2] Transactions of the Zoological Society of London, vol. iv. part vi. p. 185,

this additional evidence of the extinct bird to Mr. Broderip, and in obtaining the permission of my noble host to make such use of the painting as might best subserve the interests of Natural History. Mr. Broderip communicated to the Zoological Society the following :—

"*Notice of an Original Painting, including a Figure of the Dodo, in the Collection of His Grace the Duke of Northumberland, at Sion House.*

"Professor Owen, at whose disposal the Duke of Northumberland placed the following additional pictorial evidence of the existence of the Dodo in the seventeenth century, has requested me to draw the attention of this Society to the highly interesting picture which the Duke has been so good as to send for the inspection of the Fellows. The size of the picture, which is in the finest preservation, is thirty-two inches by nineteen. It is executed in oil, and bears the following monogram and date. Mr. William Russell, with his usual discernment, detected in this monogram the sig-

Fig. 5.

Dodo (from the painting by Goeimare, 1627, in Sion House).

natures of Jean Goeimare and Jean David de Heem, and proved the correctness of his judgment by a reference to Brulliot[1]. Jean Goeimare, who is not noticed by Descamps,

[1] Dict. des Monogrammes, 1 partie, pp. 201, 274.

Bryan, Sandrart, or Houbraken, is described by Brulliot as a Flemish artist who flourished at the commencement of the seventeenth century, and painted landscapes with many animals, executed with great care, but in rather a dry manner[1]. Of De Heem, the celebrated painter of still life, it would be superfluous to say anything. We may conclude, then, that in this joint production the landscape and animals were painted by Goeimare, and the shells by De Heem.

"In this picture, which seems to have been intended as a record of rarities, the foreground represents a sea-shore from which the tide has retired, leaving empty shells of the following genera:—*Nautilus, Pteroceras, Strombus, Triton, Pyrula, Cassis, Cyprœa, Conus, Mitra, Turbo, Nerita, Mytilus, Ostrea,* &c. Behind, on elevated ground, are two Ostriches; and below, to the right of the spectator, the Dodo is represented as in the act of picking up something from the strand" (fig. 5). "The head and body of the bird, covering an area as large as the palm of a man's hand, are seen; but the legs are hidden. The painter of the Dodo, in my picture" (Pl. I. fig. 2), "has given the only complete foreshortened back view of the bird known to me. In the Duke's picture the head and body are presented to the spectator on a larger scale; and I have nowhere seen the hood or ridge at the base of the bill, from which the bird obtained the name of *Cygnus cucullatus,* so clearly represented. Near the Dodo are a Smew and other aquatic birds, and further off Hoopoes and Terns. In the distance is the ocean, with a sea-monster awaiting the attack of Perseus, who descends on a winged steed to the rescue of Andromeda chained to a rock. Those who have had occasion to describe and figure new species of Testacea, know how difficult it is to find a draughtsman who can give a correct design of the shell to be represented. Unless the artist, like Mr. G. B. Sowerby, jun., is aware of the internal structure of the shell, and acquainted with its organization, a lamentable failure is generally the result. In the picture before us, with one exception—and even in that the specimen may have been distorted—so accurate was the eye of the painter, that if he had been aware of the organization of each shell—knowledge which he probably had not—he could not have represented the objects more correctly. The *Nautili*[2], *Strombus gigas, Triton,* and *Pyrula* are painted with great breadth and power, and all are drawn and coloured with wonderful truth; indeed a conchologist may name every species. One of the *Nautili* is partially uncoated, to show the nacre, and the other dissected, to display the concamerations. None of the shells have the epidermis, and all are of the natural size. The artificial condition of these subjects, and especially of the *Nautili*, is, it must be allowed, rather out of place in an assemblage of testaceans left on the sands by the retired tide, unless we are to suppose that the sea-nymphs had been amusing themselves by polishing the specimens and displaying the internal structure of one of them; but this very treatment shows that the designs were accurately made from real objects then considered as rarities. With the exception of the Dodo, none of the natural objects represented are now rare. The shells, especially those whose habitats

[1] "I am indebted to Mr. Russell for this information."   [2] "*Nautilus pompilius.*"

are the seas of the Antilles, are at present very common; but at the date of the picture —the second year of the reign of our first Charles—the natural productions of the West Indies were not well known, and were, comparatively, very scarce. With the shells on the shore is the cranium of a carnivorous quadruped, apparently of the family *Canidæ*. The monster-cetacean in the distance has evidently no chance with the avenger who is coming down upon him mounted on a winged steed. But Pegasus, who, with other prodigies, sprang from the blood that dropped from Medusa's head, as the conqueror who had cut it off with his harpe traversed the air with his gory trophy, immediately winged its flight to Helicon, there to become the pet of the Muses. The best version of this mythological story relates, that when Perseus afterwards killed the sea-monster and delivered Andromeda on the coast of Ethiopia, he effected his purpose by raising himself in the air through the aid of the wings and talaria given to him by Mercury, and not with the help of the winged horse on which most of the painters mount him.

"Professor Owen informs me that Roland Savery's picture containing the Dodo, in the Berlin collection, bears the date of 1626; and that the colour of the Dodo in the Duke of Northumberland's picture resembles that of the portrait of the bird, of life size, by the same painter, now at Oxford. L'Estrange describes the hue of the back of the living Dodo which he saw exhibited in London 'about 1638,' as of 'dunn or deare colour.'"

The picture of the Dodo at Berlin by R. Savery, to which Mr. Broderip refers, is copied in figure 1, Plate 1. Another figure of the bird, by the same artist, is introduced into a painting in the Imperial Collection of the Belvedere at Vienna. Fig. 3, Plate 1. of the present work, is from the copy of this picture, transmitted by Dr. Tschudi to Mr. Strickland, and given at p. 30 of the 'Dodo and its Kindred.' The date of the picture is 1628.

We have thus evidence of figures of the bird being introduced into paintings executed during the years 1626, 1627, and 1628. The different attitudes and life-like actions of the Dodo, in these representations, indicate that the artists had a living model before them. Their original studies may, indeed, have been executed at some period antecedent to the dates of the paintings into the subjects of which this rare and curious bird is introduced; but the capital fact remains, viz. that the figures given in Plate 1. faithfully represent the shape, colour, and attitudes of the now extinct brevipennate bird of the Mauritius. Different conjectures have been propounded as to the time, place, and other circumstances under which Roelandt Savery and Jean Goeimare were enabled to execute their drawings or studies of the living Dodo, and I had the satisfaction to find that Mr. Strickland concurred in the conclusion at which I arrived after my researches in Holland into the history and evidences of the bird.

"As Roland Savery was born in 1576, he was twenty-three years old when Van Neck's expedition returned to Holland, and as we are told by De Bry that the Dutch brought home a Dodo on that occasion, it is possible enough that Savery may have taken the portrait of this individual, and that the design thus made may have been copied by himself and by his nephew **John** in their later pictures. Or if we feel disposed to doubt the correctness of De Bry's statement, we may yet suppose, with Professor Owen, that the menagerie of Prince Maurice supplied the living prototype for Savery's pencil. This opinion is **corroborated by** the tradition recorded by Edwards, **that the picture in the British Museum** was drawn in Holland **from** the living bird. It is far more probable than the conjecture of Dr. Hamel (Bull. Ac. Petersb. vol. v. p. 317), that Savery's pictures were copied from the Dodo exhibited in London, as this individual must in that case have lived in captivity at least twelve years, from 1626 to 1638[1]."

With the view to test the tradition recorded by Edwards as to the date and origin of the painting of the Dodo in the British Museum, I took a copy of the outline of the bird and laid upon it outlines of the bones of the Dodo subsequently to be described, as shown in Plate III., and thus obtained proof that the painting truly represented the natural size and shape of the *Didus ineptus*, and had no doubt been "drawn in Holland from the living bird[2]." From the date of the first landing of the Dutch on the Island of Mauritius, in 1598, to their colonization of it in 1644, their ships frequently, perhaps annually, visited that island, and, as recorded by most of the writers quoted by Broderip, and testified by Van der Hagen, in 1607[3], their crews feasted on Tortoises, Dodos, Doves, and other game, and also salted the Tortoises and Dodos for consumption during the voyage to the spice-islands of the Indian Archipelago. It is highly probable that more than one of the strange birds of Prince Maurice's Island would be brought alive to Holland, and we know that a specimen was brought from that country for exhibition in London in the year 1638. It is certain that through the attacks of man, and those of the dogs, cats, and swine introduced by the Dutch into the Mauritius, the slow and heavy flightless Dodos were extirpated, probably before Leguat's visit to the island in 1693. The French colonists, who succeeded the Dutch in 1712, seem not to have found any Dodos remaining in the island; their descendants and successors have preserved no traditions of the living bird; and Baron Grant, who resided in the Mauritius from 1740 to 1760, expressly states that no such bird was to be found there at that time[4].

Mr. Broderip refers, in his History of the Dodo, to the notice by Adam Olearius, in 1666, **of the head** of that bird in the museum of the Duke of Gottorp.

This specimen was most unexpectedly discovered by Professor Reinhardt in the **Museum of Natural History at Copenhagen** under the following circumstances:—" In

---

[1] *Op. cit.* p. 30.

[2] Edwards's 'Natural History of Birds and other Rare and undescribed Animals,' &c., 4to, vol. vi, pl. 294, 1760.

[3] " Pendant **tout le temps qu'on fut là, on vécut de tortues,** de dodaros, de pigeons, de perroquets gris, et d'autre chasse, qu'on allait prendre avec les mains dans les bois. . . . . La chair des tortues terrestres était d'un fort bon goût. On en sale, et l'on fit fumer, dont on se trouve fort bien, de même que des dodaros qu'on sale." (Recueil des Voyages de la **Compagnie** des Indes Or., vol. iii. pp. 195, 196, quoted by Strickland, *op. cit.* p. 17.)

[4] ' History of the Mauritius,' p. 145[*], compiled from the Baron's papers by his son.

D

the summer of 1840 I happened to search through a box wherein different natural-history objects were stored, which had been presented by the ' Kunstkammer ' to the Royal Natural History Museum, and on this occasion I found a very large bird-cranium, which attracted my attention partly through its size, partly through its unusual and peculiar shape, and by a further examination and comparison with the authenticated representations of the Dodo, I became persuaded that it must have belonged to that remarkable bird.

" It is very well preserved, only wanting the left ' os pterygoideum ;' and the ' con-dylus occipitalis,' together with the entire border of the ' foramen magnum ' are broken away ; otherwise it is quite perfect, so that an almost complete description of the osteology of the head of this remarkable genus may be made out from it. Although I have searched through Laurentz's ' Museum Regium ' and the MS. Catalogue of the ' Kunstkammer,' I have nowhere been able to discover any notice of such a cranium having ever been possessed by the Collection, and it is therefore clear that it has pre-served the present specimen quite unwittingly, and it stands probably under one of the many numbers given as referring to heads of unknown foreign birds. I have mean-while gradually come to the conclusion that this head is in all likelihood the one called ' Dodo's head ' by Olearius in the year 1666, in his description of the Gottorp Kunst-Museum, which, when that museum, at least in part, was amalgamated with the Copenhagen Museum, found its way there." (Reinhardt, in ' Kröyer's Naturhist. Tidsskr.' iv. pp. 71, 72 (1842).

About ten years afterwards a portion of the bone of the upper beak of a Dodo was discovered in the Imperial and Royal Museum of Natural History at Prague[1].

Such, until the year 1865, was the sum of the remains of this large, flightless, extinct bird which were known to have reached Europe.

The happy perception, by the Danish Professor J. Reinhardt, in 1843[2], of the resemblance of the beak of the Dodo to that of the tropical Doves, generically separated by Cuvier under the name Vinago, on account of their proportionately larger, more strongly arched, and compressed beak than in other Pigeons, and the still closer resemblance, in miniature, of the beak of the Samoan Dove to that of the great Mau-ritian bird, which led Titian Peale to give to the former the generic name Didun-

---

[1] See Annals of Nat. Hist. ser. 2. vol. vi. p. 290 (1850).

[2] " Es war im 1843, dass ich auf den Gedanken kam, dass der Dodo eine monate Taubendors sei ; ich überzeugte mich bald dass diese Auffassung die einzig richtige sei, und fing an eine Arbeit über diesen Gegen-stand vorzubereiten. In 1845 wurde ich aber von meiner Regierung beauftragt eine Reise um die Welt mit einem dänischen Kriegsschiff mitzumachen ; meine Arbeit musste also vorläufig bei Seite gelegt werden. Schon vor meiner Abreise hat ich aber mehrere sowohl dänische wie fremde Naturforscher mit meiner Ansicht bekannt gemacht, und der Beweis das es auch so verhält wird Owen finden können :—

"1. in den Forhandlinger de Scandinaviske Naturforskere Møde, i Kjöbenhavn, 1847, p. 648 ; and

"2. in Sundevall, Arsberättelse om Framstegen i vertebrerade Djurens Naturalhistoria og Ethnographien, 1845–50, p. 234."—Letter from Prof. J. Reinhardt to Dr. Albert Günther.

culus, directed the ornithologist and ornithotomist to the family in which the most instructive comparisons might be made; and the results of these, so far as relates to to the head and foot and the bones of those parts, published by the authors of the above-cited work (p.4), left little **doubt** of the "striking affinity which exists between this extinct bird and the Pigeons".

Whatever doubt, indeed, may have lingered in the minds of naturalists as to this affinity will probably be finally set at rest by the results of the comparison of the large proportion of the skeleton of the *Didus ineptus* which has at length been transmitted from the island of Mauritius to London, under the following circumstances.

In 1863, I was favoured by Miss A. Burdett Coutts with an introduction to the Bishop of Mauritius, then in this country, and I endeavoured to interest his lordship in aiding or promoting the acquisition, by the British Museum, of the zoological rarities of Madagascar, and especially of any remains of the Dodo which might be discovered in the island of Mauritius, to which his lordship was about to return.

How speedily and successfully the Bishop has fulfilled my latter desire will be shown by the following letter, with which I was favoured in November, 1865.

"St. James, Port Louis,
"October 7, 1865.

"MY DEAR SIR,—when I had the pleasure of conversing with you for a short time in London two years ago, I promised to acquaint you with any facts or discoveries which might come to my knowledge, likely to interest you in connexion with Madagascar. I have not anything as yet to communicate definitely respecting that island in the way of natural history, but I have strong reasons to believe that a discovery has been made here recently which will gratify you very much. Mr. George Clark, who has for many years devoted himself to the work of teaching in this island with great success, is an ardent student of natural history, and has explored many parts of the island in search of information on the subject. From careful observation he was led to conclude that no remains of the Dodo were likely to be found in any of our watercourses, because of their steep descent and the immense rush of water which sweeps down them at times. But he had also frequently expressed his opinion that in certain marshes, with high banks of sand between them and the sea, such remains would probably be found. In one of these places he has found several of the bones of the Dodo (as he believes), and is now forwarding them home for your inspection [*].

At his request, I write these lines to ask for your kind care of his interests in securing any reward which may accrue to him. It would be a great pleasure to me to find that his discovery was really important, and likely to be useful to himself;

[*] Reinhardt, quoted by Strickland, *op. cit.* p. 41 (see also p. 70).
[†] This Collection was purchased by the Trustees of the British Museum for the sum of £100.

for he has pursued these and similar investigations with an amount of intelligence, skill, and diligence, in his vacation-times (by no means extensive), which deserves much credit and encouragement.

"The book which you kindly sent me on the Aye-Aye has been read by many, and especially by medical men, with much interest. I entrusted the other copy to Mr. John Douglas for the Society here.

"I remain, my dear Sir,

"Your very faithful Servant,

(Signed) "VINCENT N. MAURITIUS."

"*Professor Owen.*"

This letter was accompanied with the following "Statement" by Mr. George Clark, Master of the Government School at Mahébourg, Island of Mauritius :—

"On the estate called 'Plaisance,' about three miles from Mahébourg, in the island of Mauritius, there is a ravine of no great depth or steepness, which, apparently, once conveyed to the sea the drainings of a considerable extent of circumjacent land, but which has been stopped to seaward, most likely for ages, by an accumulation of sand extending all along the shore. The outlet from this ravine having been thus impeded, a sort of bog has been formed, called 'La Mare aux Songes,' in which is a deposit of alluvium, varying in depth, on account of the inequalities of the bottom, which is formed of large masses of basalt, from three to ten or twelve feet. The proprietor of the estate a few weeks ago conceived the idea of employing this alluvium as manure; and shortly after, the men began digging in it; when they had got to a depth of three or four feet they found numerous bones of large tortoises, among which were a carapace and a plastron pretty nearly entire, as also several crania.

"When I heard of this, it immediately struck me that the spot was one of the most likely possible to contain bones of the Dodo, and I gave directions to the men working there to look out for any bones they might find. Nothing, however, was turned up but a fragment of what I supposed to be the humerus of a large bird. This encouraged me to look further; and my search was rewarded by the discovery of several tibiæ, more or less perfect, two tarsi, one nearly perfect pelvis, and fragments of three others.

"These were found imbedded in a black vegetable mould, the lighter-coloured specimens being near the springs. My reasons for believing these to be remains of the Dodo are:—the certainty that that bird once existed in Mauritius; the similarity of these bones to what the representations of the Dodo which I have seen would lead one to expect, particularly the breadth of the pelvis, the stoutness of the tibiæ and tarsi, and the shortness of the latter; the favourable nature of the spot in which they were found for the haunts of such birds when living—a sheltered hollow with two springs in it; the non-existence, actual or traditional, in Mauritius of any bird to which bones such as these could have belonged; the indubitable antiquity of these bones, proved by the deposit of alluvium which covered them.

" During nearly thirty years that I have inhabited this colony, I have made frequent inquiries of old people as to the finding of the bones of large birds, and have offered liberal rewards for such ; and I have consulted with the late Dr. Ayres as to the spots most likely to contain them. We agreed that the floods which sweep the hill-sides and the ravines in the rainy season would be most likely to carry any remains into the sea : and this would doubtless have been the case here, but for the stoppage occasioned by the sand-down. (Signed) " GEORGE CLARK. 1865."

The above " Statement " was authenticated by the following testimony :—

" Having visited the place with Mr. Clark, I can vouch for the truth of the facts herein mentioned. (Signed) " WILLIAM THOMAS BANKS,
" Civil Chaplain, Mauritius."

"The Rev. W. T. Banks, Civil Chaplain at Mahébourg, in this diocese, and Mr. George Clark, Master of the Government School at Mahébourg, are well known to me, and deserving implicit credit for their statements as to matters of fact.

(Signed) " VINCENT N. MAURITIUS. Oct. 6, 1865."

§ 2. *Description of the Skeleton.* (Plate III.)

The bones of the Dodo (*Didus ineptus*, Linn.) discovered by Mr. Clark, under the above circumstances, which have reached me up to the present date (December 20th, 1865) are the following :—

Name.	Number of bones or parts.
Cranium and lower jaw, in parts	14
Vertebræ and pelvis	30
Ribs	22
Sternum	2
Scapular arch, in parts	7
Humerus, ulna, radius	6
Femora	5
Tibiæ	6
Fibulæ	4
Metatarsals	4
Total number of parts of skeleton of the Dodo	100

The known characters of the skull and metatarsus of the *Didus ineptus* served to identify those bones as belonging to that species : the agreement in relative size, colour, condition, and locality left no room for hesitation in referring the other bones in the above list to the same species[1]. They belong, however, to four or five individuals

[1] So determined, subsequent sets of bones transmitted from the Mauritius, and from which I was privileged to select the most perfect specimens for the present memoir, got into the market and were sold by auction since the present memoir was in type, as bones certified by me to be of the Dodo. I have to express my sincere and

varying somewhat in size. With the bones of the Dodo were the end of the lower jaw of a broad-billed Parrot, two bones (radius) of a small Mammal, and part of the skull of a large Tortoise.

To the description of the Dodo's bones I now proceed.

Vertebræ. (Plates III., IV., V., VIII., XI.)

The dorsal vertebræ are chiefly represented, in this series of bones, by three which are anchylosed together by their bodies and neural arches (Pl. V. figs. 1-5): the posterior articular surface of the body of the last **of these vertebræ** (ib., fig. 4, *c*) is subquadrate, longer in the vertical than the transverse direction, concave vertically, convex transversely, almost fitting, but being rather too small for, the anterior articular surface of the body of the first of the sacral series (Pl. VII. fig. 1, *c*). The difference is such as to indicate that only one dorsal vertebra may have intervened; and I conclude that the last of the **three** coalesced vertebræ is the penultimate dorsal. The anterior articular surface of the foremost of the three (Pl. IV. fig. 1, *c*) is 11 lines in transverse, and 4 to 5 lines in vertical diameter: it is concave transversely for the middle three-fifths, and convex transversely at the two outer fifths of its extent: it is more or less convex vertically throughout its extent. The bodies of these vertebræ are compressed

grateful acknowledgements to those gentlemen into whose hands these lots have fallen, who have forborne their own advantage and refrained from rushing into print with figures from inferior specimens to anticipate the appearance of a Memoir communicated to the Zoological Society of London, January 9th, 1866, and notified in the 'Proceedings of the Zoological Society' for January 1866 as destined "to be published entire in the Society's Transactions," and therefore necessarily awaiting the lithographing of "illustrations," which every true promoter of science for its own sake must have desired to see as complete as the best-selected materials would permit to be given.—R. O., June 1866.

¹ In the quaint print, in folio 3, of the "Narration Historique du Voiage faict par les huict Navires d'Amsterdam au mois de Mars l'An 1598. soubs la conduitte de l'admiral Jaques Corneille Neuq," &c., the first-named object, No 1, "Sont Tortues qui se tiennent sur l'haut pays, frustes d'asies pour nage, de telle grandeur, qu'ils chargent ung homme et rampent encore fort roidement, prennent aussi des Escrivisses de la grandeur d'un pied qu'ils mangent. 2. Est ung oiseau, par nous nommé Oiseau de Nausée, à l'instar d'une Cigne, ont le cul rond, couvert de deux ou trois plumettes crespues, carent des aisles, mais en lieu d'icelles ont ils trois ou quatre plumettes noires, des suedicts oiseaux avons nous pris une certaine quantité, accompaigné d'aucunes tourturelles et autres oiseaux, qui par nos compaignons furit prins, la premiere fois qu'il arrivoyent au pays, pour chercher la plus profonde et plus fraische Riviere, et si les navires y pourroyent estre sauvez, et retournerent d'une grande joye, distribuant chasque navire, de leur Venoison prins, dont nous partismes le lendemain vers le port, fournissmes chasque navire d'un Pilote de ceux qui au paravant y avoyent esté, avons cuiet cest oiseau, estoit si coriace que ne le povions assez boviller, mais l'avons mangé à demy cru. Si tout qu'arrivasmes au port, envoya le Vice-Admiral nous, avecq une certaine troupe au pays, pour trouver aucun peuple, mais n'ont trouvé personne, que des Tourturelles et autres en grande abondance, lesquels nous prismes et tuasmes, car veu qu'il n'y eust personne qui les effrais, n'avoyent ils de nous nulle crainte, tindrét lieu, se laisserent assoner. En oms y est un pays abondant en poisson et oiseaux, vtere telesmet qu'il exceda tous les autres audit voyage."—Le Second Livre de la Navigation des Indes Orientales, fol., 1601. The Tortoise and Dodo in fig. 1, p. 1, of the present work, are taken from the print, p. 3, of the **above work and edition.**

and wedged-shaped, slightly expanded at their coalesced ends, produced below into subquadrate hypapophyses in the first and second (Pl. V. fig. 1, *hy*); while this process is restricted to the fore part (ib. *hy* 3), or may be represented only by a slight anterior production of the lower edge of the wedge, in the third (ib. fig. 5, *hy* 3).

The hypapophysis of the first of the three expands at its termination (Pl. IV. fig. 1, *hy*), with the hinder angle bent back to coalesce with the front one of the next hypapophysis, which is somewhat longer, and bent forward with a similar terminal expansion: a full elliptical space is intercepted by this terminal confluence of these hypapophyses (Pl. V. figs. 1 & 5, *hy*). Each vertebra shows an elliptical articular cavity (ib. figs. 1 & 5, *p*, *p* 3) for the head of the rib, near to the anterior articular surface; the long axis of this costal surface is directed from above obliquely downward and forward. The surface of the rib's tubercle cuts obliquely the lower part of the free end of the diapophysis (Pl. IV. fig. 1, *d*).

The neural arch circumscribes a canal the anterior outlet of which (ib. fig. 1, *n*) is oval with the small end downward, 5 lines in vertical, and 3½ in transverse diameter: the sides of the neural canal slightly project inward above the lower third; the posterior outlet (Pl. V. fig. 4, *n*) is more regularly elliptical in form, and rather narrower in proportion to its vertical diameter. The neurapophysis sends off from the outer and fore part of its base a stout process, which expands and divides into zygapophyses (Pl. IV. fig. 1, *z*) and diapophyses (ib. *d*); the articular surface of the former is of a full oval shape, flat, looking obliquely upward and inward; the diapophyses extend outward and a little backward: the articular surface for the tubercle of the rib is transversely elliptical and nearly flat. The hinder part of the neurapophysis expands into the postzygapophyses: these have coalesced with the prœzygapophyses in the succeeding vertebra (Pl. V. fig. 2, *z*), as has happened also between this and the third vertebra. In the last of the three vertebræ the postzygapophyses are entire (ib. *z* 3), and show very slightly concave, oval articular surfaces, looking obliquely downward and outward (ib. fig. 4, *z*). The conjugational foramina, continuously surrounded by bone, are a full ellipse, and large, the anterior one (ib. figs. 1 & 5, *f*) being 5½ lines in vertical diameter; the second (ib. *f*) is somewhat less: these foramina are also rather larger in one of the specimens than in the other. The length of the three coalesced dorsals is the same in both, viz. 2 inches 3 lines. The neural spines have run together into a continuous ridge in fig. 1, *ns*; in fig. 5 the summit is broken off in both, leaving only the anterior angle of the foremost entire; in both this inclines forward; the hinder border of the third vertebra (fig. 1, *ns*) has the same vertical parallel as the back part of the centrum. The anterior margin of the base of the spine shows a rough surface for the attachment of ligament (Pl. IV. fig. 1, *ns*). A small foramen behind the base of each of the coalesced zygapophyses (Pl. V. fig. 2, *z z*) leads to a canal descending to the neural one, and indicates superiorly the limits of the otherwise continuously ossified neural arches.

In the series of detached vertebræ, one (Pl. V. figs. 6 & 7) indicates by its neural
spine and hypapophysis a position at the base of the neck. The centrum is barely an
inch in length; its anterior surface (ib. fig. 7, c) is narrow vertically, broad transversely;
both fore and hind surfaces indicate freedom and extent of flexure. The hypapophysis
has a broad, bituberculate base (ib. hy), but is limited in fore and aft extent to the
middle third of the under surface of the centrum; its length is shown in fig. 6, hy. The
parapophysis (fig. 7, p) is slender, and expands at both attachments, with an indication
of a terminal surface. The diapophysis (d) has a larger costal surface: it sends forward a convex ridge midway between the di- and zygapophysis (z). The neural canal
(fig. 7, n) has wider and more fully elliptical outlets than the hinder dorsal vertebræ,
in relation to the greater extent of motion at the fore part of the series. I conclude
that a free pleurapophysis (pl) existed, indicating the present to be the first of the dorsal
series, as shown in Pl. III. The neural spine is short, broad, obtusely pointed, with a
vertically oblong syndesmotic surface (fig. 7) before and behind. Each postzygapophysis
(fig. 6, z') supports an anapophysial tubercle (a).

A cervical vertebra from a position just in advance of the above has lost the neural
spine, but retains the hypapophysis. This process (ib. figs. 8 & 9, hy) is compressed
and directed obliquely downward and forward for an extent of 6 lines; the extremity is
rounded: the length of the centrum of this vertebra is 1 inch 3 lines; the anterior
articular surface is longest transversely, and concave in that direction, convex vertically;
the proportions and curvatures are transposed in the posterior surface (fig. 9, c). The
parapophysis (ib. p) is continued from the anterior border of the centrum to the
middle; it is a depressed plate, confluent with the rib (ib. d). The diapophysis
forms a short, obtuse projection above its anchylosis with the rib (ib. pl): this
projects backward 7 lines in length, terminating obtusely, and circumscribing a ver-
tebrarterial foramen (ib. v) of a full elliptic shape, 5½ lines in long diameter. The
surfaces of the prozygapophyses (z) are larger, and look more upward and less inward,
than in the preceding and the dorsal vertebræ: they are very slightly concave. Those
of the postzygapophyses (fig. 8, z'), with a downward and slightly outward aspect, are
in a similar degree convex. The neural canal, as usual in the cervical series, expands
at its outlets, most so posteriorly (fig. 9, n); the middle of the upper surface of the
neural arch is impressed by an elliptical, rough, ligamentous surface, which slightly
rising in the middle is the sole indication of a neural spine. The upper surface of
each postzygapophysis developes a tuberous anapophysis (figs. 8 & 9, a).

The three cervicals that succeed the axis show progressively sinking neural spines,
which subside in the six following vertebræ (Pl. III.). The third cervical has also the
hypapophysis (Pl. XI. fig. 3, hy).

In all the other cervicals of the present series the hypapophysis is wanting, but each
parapophysis developes a plate (Pl. V. figs. 10 & 11, Pl. VIII. fig. 1, p) to form the
sides of the hæmal canal through which the carotids ran; and the position of such ver-

tebræ in the cervical series is indicated, respectively, by the degree of convergence of these processes, in none of which, where entire, have they met so as to circumscribe the canal: in some of these vertebræ, however, they are mutilated. They differ chiefly in the position and shape of the anapophyses (fig. 10, a), which advance from above the postzygapophyses (z'), converging towards the middle of the upper surface of the neural arch, being arrested, save in one instance, at the sides of the ligamentous surface occupying the common position of the base of the neural spine.

In the axis vertebra (Pl. V. figs. 12 & 13) the posterior articular surface, concave vertically, and 3 lines in that extent at its middle part, is very convex transversely, being continued upon the sides of the posterior part of the centrum; a thick obtuse hypapophysis (fig. 13, hy) descends below this surface: the anterior or odontoid surface presents the usual form in birds; the odontoid process (ib. x) has a pit at its apex. The prezygapophyses (fig. 12, z), of very small size, project from the outer and fore border of the neural arch, with their articular surface looking outward and slightly upward; a ridge is continued from their back part to the base of the postzygapophyses: the surface (fig. 13, z') in these, 4½ lines in long diameter, is three times the size of the anterior one: it is concave transversely, and looks downward and a little outward. The anapophyses (ib. fig. 12, a) are large tubercles rising above the articular surfaces. The base of the neural spine, 9 lines in length (ib. ns), is coextensive with the neural arch; the spine rises posteriorly to a height of 6 lines, with a thickness of 2 lines, having a convex upper margin (Pl. III.).

The relative size and position of the cervical vertebræ, as coadjusted in the position and degree of flexure of the neck represented in Sir Hans Sloane's life-size painting of the Dodo, in the British Museum, are given in Plate III. with the varying proportions of the pleurapophyses and other processes.

Ribs. (Plates III. & IV.)

The specimens of ribs include both vertebral and sternal portions; that which appears to be the second or third on the right side (Pl. IV. figs. 7, 7 a) is 4 inches 4 lines in length (following the outer curve), and expands to a breadth of 7 lines at its lower part; the interval between the articular surfaces of the head and tubercle is 6 lines. The appendage (ib. a) has coalesced with the middle of the hind margin of the shaft. The neck is compressed, with a thin upper margin; the lower one is continued with a curve upon a strong internal buttress-like ridge (ib. b), which runs to near the fore part of the flattened body of the rib, where it meets the ridge continued from the tubercle, about 2 inches down the rib: there is a shallow channel between these ridges, contracting to their confluence. The inner surface of the rib is impressed by a deeper and broader channel behind the buttress: the posterior border expands in in the form of a triangular plate, with a base of about an inch in extent, due to the complete confluence there of the epipleural process. The anterior border is thicker,

E

and is almost straight. Towards the sternal end the pleurapophysis contracts and thickens, terminating in a rough syndesmotic elliptical surface, 3 lines by 2 (fig. 7, *f*), for the attachment of the hæmapophysis or sternal rib.

A vertebral rib (ib. fig. 2) which is entire, measures 9 inches in length (following the outer curve). The head and tubercle are at the same distance as in the preceding, but the tubercle is broader. The characters of the body of the rib are very similar; but it is narrower, not attaining a breadth of 5½ lines at its lower end; the narrowing and thickening to the articular surface for the sternal rib is more gradual.

A last vertebral rib is adapted, by the longitudinal extent and partial division of the tubercle, to the vertebra which forms the first of the coalesced series of sacrals ; and the body of the rib, instead of preserving the regular outward curve of the antecedent ones, is more suddenly bent soon after it emerges beyond the margin of the ilium ; the lamelliform part thence continued is straighter, oval, moreover, shows upon its outer surface a flattened facet, indicative of pressure or friction by the movements to and fro of the thigh over a rib in such position. Beyond this surface the rib curves in a way not shown in the other specimens ; the distal end has the flat syndesmotic articular surface to which had been attached a hæmapophysis not reaching the sternum. In this last (eighth) free rib there is no epipleural process, nor any definitely marked ligamental surface on the posterior margin indicative of the attachment of such process.

The body of a posterior vertebral rib (Pl. IV. fig. 10) shows a fracture which has been healed, with some irregular ossific deposit on the inner surface. All the ribs have a pneumatic foramen (ib. figs. 2, 7, 8, *p*) at the fore part of the neck, near the base of the tubercle.

The eight left vertebral ribs (Pl. III.) and the five right ones do not, either of them, constitute a consecutive series, but have come from different individuals, of different sizes, as exemplified in the third rib figured in Plates III. and IV.

The sternal ribs (Pl. IV. figs. 3 & 12) are characterized by the two facets, nearly or quite meeting at an open angle, into which their sternal end expands (ib. fig. 3, *c*). One of these ribs, which is entire, shows the single, elliptic syndesmotic surface at the opposite end (ib. *b*); it is 3½ inches in length, with a greatest breadth of 5 lines, and is straight. Another and longer specimen (ib. 12) shows a moderate degree of curvature. A third specimen is 6 inches in length: the proximal end has a breadth of nearly half an inch (the penultimate rib in Pl. III.).

Five successive sternal ribs are indicated by gradational size and curvature, and a sixth, which does not reach the sternum. Before describing this bone I shall proceed with the account of the sacral vertebræ, and the expanded hæmal arches of such as complete the pelvis.

Pelvis. (Plates III. & VII.)

The pelvis of the Dodo is chiefly remarkable for the flatness and great breadth of the posterior half, corresponding with the characteristic proportions of that part of the body in Pl. I. fig. 2, and in the old woodcuts of the Dutch " Dodaersen"[1]. It includes sixteen coalesced sacral vertebræ, with which the iliac bones are continuously confluent.

The first sacral shows the transversely extended and concave articular surface of the centrum (Pl. VII. fig. 1, *c*); the subcircular pit (ib. *p*) for the head of the rib is behind the middle of the side of the centrum, at its upper part; the inferior surface is ridged lengthwise; and a transverse low but sharp ridge defines the posterior boundary, the depressions in front of which indicate the hindmost origins of the subvertebral muscle (longus colli?). The anterior outlet of the neural canal (ib. *n*) is subcircular in one specimen, vertically elliptic in others, and 3 lines or less in transverse diameter. From the sides of the neurapophyses stretch out the strong buttresses of bone which blend with the under part of the ilia, giving off from the fore part of their base the præzygapophyses (ib. *z*), and from the back part of their apex the surface (ib. *d*), or part of it, for the tubercle of the last moveable rib, the ilium in the latter variety affording the rest of that surface. The fore part of the strong neural spine (ib. *ns*) is roughened by a syndesmotic surface; it rises to a height of 14 lines, curving forward, and is confluent at its summit with the approximated anterior margins of the ilia. A continuous track of bone, forming a smoothly obtuse longitudinal ridge, represents the summits of the succeeding sacral spines (ib. fig. 2, *ns*) to the hindmost vertebra of the series, without any trace of their primitive division; but this track rises, posteriorly, above the shallow channel on each side, in which are the foramina (ib. *o*), indicating most of the constituent vertebræ.

The second sacral vertebra abuts against the ilium by a pleurapophysis (ib. fig. 1, *pl 2*), as well as a diapophysis (ib. *d 2*); but the former is a slender, straight filament, or narrow plate of bone, confluent at both ends.

In the next two vertebræ the pleurapophysis (ib. *pl 3 & 4*) assumes more breadth and robustness, but is short and straight, abutting against the inner surface of the ilium an inch in advance of the acetabulum. The first of these rib-buttresses inclines forward, and is completely confluent with the ilium; the thicker one (ib. *pl 4*) has retained part of its primitive ligamentous attachment to the ilium: the proportions of both are subject to some variety.

These are succeeded by three or four vertebræ in which the pleurapophysis is not developed, the attachment to the ilia being by diapophyses only (ib. *d d*), which are short slender lamellæ, directed upward and backward; below and between them are the double orifices for the separate motory and sensory roots of the sacro-spinal nerves. In the next vertebra the pleurapophysis (ib. *pl 8*) reappears, longer but more slender than in the fourth sacral, extending obliquely backward, and expanding at its extremity to abut against a prominence on the underside of the ilium, opposite the hind part of the

[1] See, especially, Bontius's figure, copied by Strickland, in the title-page and at p. 63 of the above-cited work.

acetabulum, with which prominence the rib has completely coalesced by an expanded end. The under part of all these vertebræ is traversed by a sharp median longitudinal ridge, which is more feebly and interruptedly continued to near the end of the sacral series.

Eight vertebræ, abutting by diapophyses only (Pl. VII. d d) against the ilia, succeed the one last described; their coalesced bodies are less than half the breadth of those of the preceding vertebræ: they gradually diminish in depth to the last, without loss of breadth. The diapophyses proceed obliquely outward and backward, are lamelliform, about 9 lines in length, and intercept oblong cavities of the same extent and direction, into which open the orifices (ib. fig. 2, o) noticed on the upper surface of that part of the pelvis. The articular surface of the body of the last sacral is transversely elliptic, 4 lines by 2 lines, and very slightly convex. The outlet of the neural canal, above it, is circular, and about a line in diameter, the whole vertical extent of the last sacral being 5 lines, while that of the first sacral is 2 inches 2 lines.

The ilium is divided, as usual, into two parts by the ridge on its upper or outer surface (ib. fig. 2, r), extending obliquely backward to behind the acetabulum—the anterior division being narrower and concave, the posterior broader and convex but in a minor degree. The anterior (slightly thickened) border of the ilium is curved with the convexity forward, extending 8 or 9 lines in advance of the fore part of the neural spine of the first sacral vertebra. The ilia almost meet above that of the second and third sacrals, with which they coalesce, and then diverge to the oblique boundary ridge, which is thence continued, in some with an angular bend, more directly outward. At this angle the bone is so confluent with the sacrum that the orifices leading to the ilconeural canals' are almost or quite obliterated. These canals are, here (ib. i i), the longitudinally extended cavities intercepted between the fore parts of the ilia and the continuous coalesced sacral spines and diapophyses, widening at their anterior outlets. The extent of that part of the ilium in advance of the acetabulum is 3 inches 8 lines; the breadth at its middle part is 2 inches. As the ilium approaches the acetabulum it increases in thickness, and is grooved at the outer margin by a vessel which leaves impressions of its ramifications upon the upper concave surface of the bone (ib. fig. 2, w). The acetabulum (ib. a a) is circular, 11 lines in the diameter of its outlet, 9 or 10 lines in that of its inner circumference, being widely open, as usual in birds, towards the cavity of the pelvis; the trochanterian surface (ib. t t) above the acetabulum is elliptic, with the long axis lengthwise, 9 lines by 6 in its diameter, with its upper border sharp and produced; the anterior border (ib. b) of the acetabulum is slightly produced; the position of this articular cavity is about midway between the fore and hind ends of the pelvis. The oblique external ridge of the ilium terminates in the outer margin of the broader part of the bone (ib. r'), 7 lines above the sharp and prominent margin of the trochanterian

' Owen, 'Anatomy of Vertebrates,' 1866, vol. ii. p. 32.

29

surface (ib. t). The ilia have diverged from each other for the extent of an inch and a **half behind the** beginning of the boundary line (ib. r), which interval is occupied exteriorly by lateral ossification from the neural spines to the diapophyses of that part **of** the sacrum : the mesial borders of the ilia (ib. fig. 2, c) slightly converge to the fifteenth sacral vertebra, where they are separated by an interspace of 1 inch, and **then** again diverge to the last sacral ; they coalesce with the diapophyses (ib. fig. 2, d d). The inner or under surface of the ilium is thickened into a kind of buttress (ib. fig. 1, e), terminating behind the ischiadic foramen. The breadth of the iliac bones and intervening sacrals, 1 inch behind the acetabulum, is 5 inches; at the back part of the pelvis it is 4 inches. The outer border of the posterior part of the ilium (ib. fig. 2, g) projects as an obtuse ridge above the ischiadic foramen and the succeeding expanded and confluent part of the ischium (ib. ca), which is vertically concave externally : the ilium, ischium, and pubis (ib. fig. 1, ca) have completely coalesced around the acetabulum. The pubis, which in this part is 7 lines thick, contracts as it becomes free to a diameter of 4 lines; it is smooth and convex below, and has been broken off near the acetabulum on both sides ; the fracture shows its pneumatic structure. The ischium, as it recedes from the acetabulum, contracts to a trihedral column, with a vertical diameter of 4 lines; it is concave outwardly, convex inwardly, and suddenly expands below, about an inch from the acetabulum, to form part of the posterior boundary of the obturator foramen (ib. fig. 1, f), which is 9 lines in length, and is situated one half in advance of, and the other half beneath, the ischiadic foramen (ib. m). This latter is oval, with the large end forwards, 1 inch 3 lines by 10 lines in its principal diameters. Behind this foramen the ischium is confluent with the ilium for an extent of 2 inches, or perhaps rather more, as the posterior margin of the pelvis is not entire in any of my specimens. The inner surface of the ischium forms a low, obtuse longitudinal ridge towards the pelvic cavity, losing thickness as it recedes from the acetabulum. The chief pneumatic foramina in the pelvis are on the inner **surface, above the acetabulum,** behind the trochanterian articulations, and behind the iliac confluence of the last sacral pleurapophyses,—also at the hinder part of the ilium, on each side of the transverse buttress (ib. e) near its posterior junction with the ischium. The prærenal fossa (between **pl** 4 & **pl** 5, fig. 1) is deep and subdivided by the diapophysial plates : the post-**renal fossa is wide and** shallow.

Sternum. (Plates III., IV., VI., XI.)

Of this instructive **and** determinative bone there are two specimens, the one most entire (Pls. III., IV. fig. **4, &** VI.) measuring in a straight line, from the costal process to the hind border, 7 inches. The extreme breadth between the lateral processes (Pl. IV. h) is 4½ inches; **from** this diameter the bone contracts anteriorly to a **breadth of 3¼ inches** at the costal processes (ib. d), and posteriorly it contracts more rapidly to an obtuse, horizontally flattened apex (Pl. VI. fig. 3). The anterior border of the sternum (Pl. IV. fig. 4) is widely and rather deeply emarginate at the middle (e), less deeply so on each side : the breadth of the mid notch (b e b) is 1 inch 9 lines, that of each side notch (b d) is 1 inch 2 lines. The sternum is deeply hollowed above (Pl. XI. fig. 4), correspondingly convex beneath (ib.); the keel (s) is low and thick, commencing by a pair of broad obtuse ridges (Pls. IV. fig. 4, & VI. fig. 1, r r) from the mesial ends of the outer walls of the coracoid grooves (ib. b´), which gradually rise from the surface of the bone as they extend backward, converging to form the beginning of the keel about 2 inches from the anterior emargination (e) : the keel gains a depth of ½ of an inch at the middle of the sternum, then gradually sinks to the level of the bone, as it extends backward, at 1½ inch from the hind end (Pl. VI. fig. 3), a little increasing in thickness as it subsides : its free border describes a pretty regular convex curve (Pl. III.) ; it is thick, flat, partially canaliculate ; the sides of the base of the keel expand, to be continued gradually into the body of the sternum (Pl. XI. fig. 4). Behind the costal surface (Pl. VI. c), on each side, extends a lamelliform process (Pls. III. & VI. b), ½ an inch in breadth, upward and a little outward, slightly expanding to its free termination, which, however, is not entire in either specimen : the longitudinal extent of this characteristic process, where it is best preserved, is 1 inch ; it is conjecturally restored in Plate III. ; it answers to the ectolateral process (h) of the gallinaceous sternum (Pls. III. & XII. fig. 3) : there is no trace of an entolateral process (ib. i). The thin margin of the Dodo's breastbone, behind the ectolateral process (Pls. III. & VI. h), is entire and uninterrupted to the obtuse apex, and the body of the sternum is imperforate : the notch (f) behind the process (h) represents the ectolateral notch of the gallinaceous sternum (Pl. XII. figs. 1 & 3, f). The costal border (Pl. VI. fig. 2, c) is 1 inch 9 lines in extent, and 6 lines across its broadest part ; it shows articular surfaces for five sternal ribs, of which the four posterior (2–5) are bilobed, the anterior one (c 1) simple, and limited to the outer half of the border ; the second sternum shows some variety in this respect : the deep interspaces, in both, are perforated by pneumatic foramina. The costal process (d)´ in advance of these surfaces expands, as it rises upward and a little outward and forward, to the extent of nearly an inch ; the hinder and outer side is impressed by a concavity, continued from the costal border ; the inner side is smooth and convex : it is not quite entire on either side. The coracoid grooves (Pl. IV. fig. 4, b b´) are small in proportion to the sternum, and are divided from each other by an interspace of about an inch ; the outer wall of the groove (b´), 9 lines in extent, is moderately produced and convex ; it appears to be a continuation of one of the initial ridges (r) of the keel : the inner wall of the groove (b) is deeper, and is formed by the obtuse angle of the anterior border of the sternum, between the medial and lateral emarginations. External to each coracoid groove is a large elliptical pneumatic foramen (p) or depression. There is no episternal process. On the convex outer surface of the body of the sternum the " pec-

---

[1] Called " hyosternal " in the Geoffroyan determination of parts of the bird's sternum.

toral" ridge (Pl. VI. fig. 1, k)' is feebly indicated, extending from the outer end of the coracoid groove backward and inward to near the posterior third of the keel. The concave surface of the sternum (ib. fig. 2) shows a number of small pneumatic foramina, chiefly along the middle line to near the posterior third. Behind the costal border the substance of the sternum gradually increases in thickness from the sharp lateral margins **to the middle, above** the **base** of **the keel, and shows there a** fine pneumocancellous texture (Pl. XI. fig. 4).

Scapular Arch. (Plates III. & VIII.)

This consists of the scapula (Pl. VIII. figs. **6, 7, 8 & 9, u), coracoid (ib.** figs. 4 & 5, u), and clavicle (ib. sc), the latter ending in a point and **here** tied by ligament to its fellow, to form **a furculum.** I have received the elements of this arch in three conditions:—one in which the bones, though of full size, are separate; **a second,** in which the scapula and coracoid are confluent, but the clavicle distinct; a third, in **which** the three bones are confluent at the ends converging to the humeral articulation. The scapula (ib. figs. 6, 7, 8 & **9,** sl), 3 inches 7 or 8 lines in length, has the usual sabre-shaped body, slightly expanding and decurved at its free extremity, the breadth of which is 7 lines: it terminates obtusely: varieties of shape are shown **in figures 6 & 8. The outer surface of the bone, at the two posterior thirds of its extent, is slightly concave and marked by muscular attachments; the inner surface of that part is smooth and slightly convex:** the bone increases in breadth, with some diminution of thickness, towards the **articular end, and is** remarkable for sending off from **the lower border, at 7 or 8 lines from that end, a short** process (ib. **sl);** between this process and the articulation the breadth **of the bone is** little more than **3 lines;** the breadth of the articular end is **9 lines. Nearly** one-half of it is occupied by **the** almost flat, subcircular humeral surface (fig. 8, a), with a diameter of 4½ lines, and directed upward, outward, and a little forward. From this is continued an oblong, much narrower coracoidal surface, beyond which the acromial process (fig. 6, c) extends forward, curving toward the coracoid, and terminating obtusely.

The coracoid (ib. figs. 4, 5, 8 & 9, sc), averaging a length of 3 inches 7 lines, expands **to a breadth of 1** inch 3 lines at its sternal end (sc), of which the articular surface (e) **occupies an inch;** the non-articular part forms the **outer** angle (m), and extends in advance **of the pneumatic foramen** (Pl. IV. fig. 4, p) at that part **of the breast-bone: the outer border which extends from this free angle to the** body of the bone, into which it subsides, at one-third of the extent of **the bone, is sharp;** the inner border is obtuse to near the inner angle (Pl. VIII. figs. **4 & 5, n). The outer surface of** the expanded sternal end is smooth and convex; the inner surface is flatter and more irregular, perforated by pneumatic foramina; the diameter of the subcylindrical part of the shaft is 4 lines: the **extremes** of difference in the distal expansion of the coracoid are shown in figs. 4 & 8, sc,

---

' The intermuscular ridges ('pectoral,' 'subcostal,' '**carinal**') are, with other parts of the bird's sternum, **here named as defined in** my 'Anatomy of Vertebrates,' vol. ii. pp. 16-23.

Pl. VIII. A muscular ridge and rough surface (ib. fig. 9, *r*) mark the back part below the middle of the shaft. The bone then expands to its upper articular end, which is obliquely truncate from within outward: it shows, first, the oblong surface for the scapula, which is extended upon the inner prominence of that end; next, the larger and full oval surface for the humerus (*h*), from which the thick, obtuse, inner continuation of the scapular end projects inward, forward, with a slightly upward curve, and shows the narrow oblong surface for the articulation and ultimate confluence of the clavicle (*s*). The coracoid unites with the scapula at an angle of 100°.

The clavicle (ib. figs. 4 & 5, *ss*), at its scapular end, is slightly expanded, compressed, with an obtuse recurved termination articulating with the above-named surface of the coracoid, and in one instance coalescing therewith, and by extended ossification with the "acromion scapulæ" (ib. figs. 8 & 9). As the clavicle descends it curves slightly and contracts to a point. The angle at which the pair meet is shown in figs. 4 & 5.

Bones of the Wing. (Pls. III. & VIII. figs. 12–17.)

Of the humerus the series contains two specimens, both measuring 4 inches 3 lines in length, one right, and the other left (Pl. VIII. figs. 12–14), but differing slightly in their proportions and in colour—one being of the olive-brown tint with which most of the bones are stained, the other black. The articular head (ib. *a*) is an elongate oval convexity, with the larger end toward the radial side, prominent toward the back and rather flattened toward the front of the bone, which there swells out beyond the base of the articular surface. The radial tubercle is small, and descends from the radial end of the head for about 5 lines; the pectoral process (ib. *b*) is triangular, obtuse, short, and bent, or directed toward the front side of the bone; the ulnar tuberosity (ib. *c*) is more produced in that direction; it is oblong, obtuse, with its base impressed by a large pit both above (fig. 12, *h*) and below—the lower one (ib. *g*) being the deepest, and perforated by a pneumatic foramen; the convex, broad, ulnar border of this tuberosity has two slightly produced processes, an upper or posterior (ib. fig. 12, *c*) and a lower and internal (ib. *g*), which is the smallest. The breadth of the proximal end of the humerus, across the tuberosities, is 1 inch 5 lines, beyond them the bone contracts to a smooth subcylindrical shaft, showing at the back part of the proximal third a longitudinal ridge (fig. 12, *r*), half an inch in length; it gradually expands at the distal third to a breadth of 10 lines, where the articulations offer the usual avian characteristics of the elbow-joint. The head of the humerus is occupied by a fine cancellous structure: into the large vacuity below this, crossed in the section figured (Pl. XI. fig. 5) by a transverse slender bar of bone, the small pneumatic foramina at the bottom of the wide and deep fossa for the axillary air-cell open. The part of the hollow proximal end giving off the pectoral and other processes for the attachment of muscles is strengthened by similar abutments. The pneumatic cavity of the main part of the shaft of the humerus is simple, with a compact wall thicker than at the ends of the humerus, but not exceeding that which is characteristic of the long air-bones in birds. The portion of the distal end chiefly serving for muscular attachments and the antibrachial articulation are also cancellous.

The *radius* (Pls. III. & XII. fig. 15) is a straight and slender bone, 3 inches 1 line in in length, and 2 lines in chief diameter of the shaft. The proximal articular surface is subcircular, 3 lines in diameter, moderately concave ; the distal end expands to the same extent, but is compressed, as usual.

The *ulna* (Pls. III. & VIII. figs. 16 & 17) is 3 inches 1 line in length, of the usual ornithic character, with a well-defined, narrow, elliptic, rough muscular depression, 8 lines in length (fig. 16, *c*), extending upon the shaft from below the anterior or palmar angle of the proximal articular surface. This bone has no pneumatic foramen ; the orifice for the medullary artery is above the middle of the same palmar surface, the canal inclining distad. The shaft of the bone is nearly straight ; the back or anconal surface, which is slightly convex, shows feeble impressions of the attaching ligaments of the alar plumes, which are represented in all the figures of the entire or living bird. A second ulna is 3 inches 3 lines in length.

There was no carpal or pinion bone in the collection of remains submitted to me : this part of the wing is conjecturally restored in dotted outline in Plate XV.

Bones of the Leg. (Pls. III., IX., X. & XI.)

Of the five *femora* in the above defined series of remains of the Dodo, two measure 6 inches 3 lines in length ; one (Pl. IX.) is 6 inches 4½ lines ; the shortest is a little under 6 inches, with proportionate differences in the diameter of the shaft. All of them show a small pneumatic foramen (Pl. IX. figs. 1 & 2, *p*) on the inner side of the anterior ridge of the great trochanter (ib. *c*), and on the same transverse line with the head of the bone. This part shows an oblong depression (ib. figs. 2 & 3, *a*) for the " ligamentum teres" at the upper and back part. The articular surface on the same aspect of the neck (ib. fig. 3, *b*), adapted to the trochanterian prominence of the pelvis (Pl. VII. *t*), is well-defined. The trochanter (Pl. IX. fig. 1, *c*) rises, ridge-like, above the level of the head, and is continued from behind the middle of the articular surface on the neck, forward, with a convex outline upon the fore and outer part of the shaft, where it gradually subsides ; a narrow intermuscular ridge (ib. fig. 1, *r*), inclining to the middle of the fore part of the shaft, is continued from the trochanterian one. The small trochanter (ib. fig. 3, *d*) is a small subcircular tuberosity, in some specimens a ridge, 3 to 4 lines in length, on the inner side of the shaft, about an inch below the head. The muscular impressions on the fore part of the bone are well defined. A minute medullary canal (ib. fig. 3, *m*) perforates the middle of the back part of the shaft ; the popliteal fossa (ib. fig. 3, *o*) shows a few small pneumatic orifices ; a triangular rough flat surface divides the fossa from the outer condyle. Above the fibular depression (ib. fig. 3, *g*) there is a well-defined, slightly raised, rough surface (ib. *k*) for the head of the ectogastrocnemius muscle. The ridge (ib. *n*) extending to the back part of the

F

inner condyle is not sharp; the rotular groove (ib. fig. 1, *p*) is deep and moderately wide, with the inner boundary, formed by the narrow anterior part of the inner condyle (ib. fig. 5, *c'*), most produced. The breadth of this end of the longer femora is 1 inch 9 lines; the character of the distal articular surface is shown in Pl. IX. fig. 5.

The head, neck, and great trochanter (Pl. XI. fig. 6) are occupied by a pneumatic cancellous structure, with a thin compact wall on the upper part and sides; this begins to gain thickness at the under part of the neck and at the lower and back part of the trochanter, the compact wall acquiring a thickness of a line at the beginning of the shaft, where the cancellous structure is confined to the outer side of the pneumatic cavity; this structure gives way to a few delicate filaments of bone crossing the cavity of the major part of the shaft, and is not resumed until the bone expands to form the distal condyles (ib. fig. 7).

The five *tibiæ* of *Didus* in the same collection range in length from 8 inches 8 lines to 9 inches. The procnemial ridge (Pl. X. figs. 1, 2, 4, *p*) is a triangular plate, with the base longest and the apex rounded off; it inclines outwardly, and does not extend much more than half an inch from the level of the proximal end of the bone; the length of its base rather exceeds an inch: on its inner side a triangular muscular surface is well defined by an irregular inferior line or ridge (ib. fig. 2, *n*). The ectocnemial process (ib. figs. 1, 3, 4, *e*) is thicker, shorter, and terminates roughly and obtusely. There is a low, narrow ridge (ib. fig. 2, *g*), about half an inch in length, on the inner side of the proximal end of the shaft, beginning about 9 lines below the articular surface at that end. The fibular ridge (ib. figs. 1 & 3, *h*), beginning 1 inch 8 lines from the proximal end, extends about 2 inches down the outer side of the shaft. The epicnemial ridge (ib. figs. 1 & 4, *k*) is obtuse, and but little produced above the upper articular surfaces or condyles (*t d*) of the tibia: the breadth of that end of the bone, in the longest specimen, is 2 inches 3 lines. The tendinal canal at the fore part of the distal end is bridged by bone (ib. fig. 1, *i*), and is situated on the inner half of that aspect of the shaft; the lower opening is subcircular and close to the anterior end of the inner lower condyle (ib. *o*), which is more produced forward than the outer one (ib. *o*). Their hind ends project very little beyond the level of that aspect of the shaft of the tibia. An intermuscular ridge (ib. fig. 1, *z*) strengthens into a tuberosity (*r*) at the inner side of the tendinal groove.

The cancellous structure in the tibia is limited to an extent of about half an inch below the proximal articular surfaces (Pl. XI. fig. 8), and to about an inch and a half from the distal end of the line (ib. fig. 9): the shaft is occupied by a large air-cavity, with a compact wall of half a line in thickness at the upper third, gradually increasing to about a line at the lower fourth, until the cancellous structure is reestablished; the transverse direction of a plate of this structure indicates the extent of the original distal epiphysis of the tibia (fig. 8).

The *fibula* (Pl. X. figs. 6-8) presents the usual ornithic characters of the bone:

it varies from 4 inches 4 lines to 4 inches 6 lines in length, with a greatest proximal
breadth of 8 lines. No adequate gain would result from a detailed description or comparison of this bone; and the rest of the bones of the foot have received every requisite
attention in this way in the excellent work on the Dodo and its kindred, already
quoted. A longitudinal section of the *metatarsus*, taken in the direction from side to
side (Pl. XI. fig. 10), shows the loose cancellous texture of the common epiphysis of
the three long metatarsals, and the remnant of their contiguous coalesced walls reduced
to a thin lamella of bone. As the moiety of the bone figured is the posterior one (of
the left metatarsus), the usual oblique position of the middle metatarsal (*iii*), with its
proximal end nearer the back part and its distal end nearer the fore part of the coalesced
series, produces a corresponding direction of the section, with narrowing and termination
of the exposed part of the medullary canal about one-third from the distal end of that
metatarsal. The medullary canal of the outer metatarsal (*iv*) is wider and descends
lower before the breaking up of the inner surface into decussating lamellæ or filaments,
than that of the inner metatarsal (*ii*): the peripheral compact wall of the inner is twice
the thickness of that of the outer metatarsal. I may remark that the more posterior
position of the middle metatarsal at its proximal end, from which and the corresponding
part of the common epiphysis the calcaneal process is developed, is related to the greater
share taken by the middle toe in the act of walking and scratching. I will only remark
that of the four metatarsals of as many Dodos in the present series, one exceeds by a
line the length of that figured in plate xi. *op. cit.*, and one falls short thereof to the
same trifling amount.

Skull. (Plates III. & XI. fig. 1.)

Of the skull of the Dodo, the series of bones transmitted to me include the cranial
part with the detached upper mandibular bone (more or less mutilated) of two mature
birds, and the lower mandible of three individuals. In the latter the dentary elements
(Pl. XI. fig. 1, *a*), confluent at the "gonys," are distinct from the hinder halves of
the rami formed by the confluent, or perhaps connate, articular, surangular and angular elements (ib. *a*): if the "splenial" were ever distinct, it has coalesced with the
dentary, where its upper boundary is indicated by a linear groove or series of small
foramina.

In size, shape, and all other characters of these important evidences of the specific
nature of the remains from the Mahébourg morass[1], they agree with those of *Didus
ineptus* detailed in the 'Proceedings of the Zoological Society' for January 11th, 1848
(part xvi. pp. 2–8), and in the work entitled "The Dodo and its Kindred," pp. 76–96.

The occipital condyle (ib. *i*) presents the same hemispheroid or reniform shape, with
the median vertical notch or depression above. The upper margin of the foramen magnum is broad, as it were excised, with the sides slightly prominent. The superoccipital

[1] "La Mare aux Songes."

F 2

foramen is present in both specimens, as in the one originally described (Proc. Zool. Soc. part xvi. p. 2). This foramen also exists in Owls and Parrots, but not in all Pigeons; the *Didunculus* (Pl. III. fig. 2) shows no trace of it; I have also failed to find it in the skull of a Crown-pigeon (*Goura coronata*). The superoccipital ridge is defined by the subsidence of the surface beneath it being continued directly from the upper, almost flat, smooth surface of the cranium: the middle part of the ridge is more produced than the angles. In the great breadth of the occipital surface compared with its depth, in its flatness from side to side, and its aspect backward and a little upward, *Didus* most resembles *Dinornis*. The basioccipital curves downward, and unites with the basi-sphenoid in developing the pair of larger tuberosities (Pl. XI. fig. 1, 3), which terminate about ½ an inch below the occipital condyle. There is nothing of this structure in the Columbine cranium. In one of my Dodo's skulls there is a pair of small tubercles between the larger basioccipital ones; these are not developed in the other cranium. The basisphenoid is subquadrate, and flattish below, impressed by a shallow median longitudinal channel.

The hypoglossal nerve escapes by two small foramina on each side of the base of the condyle; external to these is the vagal foramen; still more external is the depression (ib. *a*) perforated below by the entocarotid, glossopharyngeal, and sympathetic, above by the tympanic vein. The entocarotid canal opens into the hind part of the sella or pituitary fossa: the vagal canal begins within the skull, above the hypoglossal foramina. The paroccipital carries the posterior surface of the skull downward and outward to a much greater degree than in any Dove, but to a less degree than in *Dinornis*. The Eustachian tubes impress the outer and fore part of the basisphenoid.

The temporal fossæ (Pl. III.), in the present specimens, show the same contraction in proportion to their depth by which the original skull of the Dodo, compared with that of the *Dinornis*, ' Proc. Zool. Soc.' (1848, p. 3), differed from the larger extinct wingless bird. In the approximation of the postorbital process to the mastoid, *Didunculus* shows a closer resemblance to *Didus* than does *Goura*, in which the temporal fossa, besides being narrow, is shallow. The temporal muscle appears to spread its origin above the fossa upon the sides of the cranium, forward half an inch in advance of the postfrontal process, and backward to the outer angle of the superoccipital ridge.

The parietal region is broad, flat, and short, as in *Dinornis*, not convex as in Doves; it is also impressed at its middle part by a shallow transverse groove, continued outward and forward of less depth and definition, so as to mark off the convex interorbital part of the swollen frontals.

The outer side of the mastoid is convex, smooth, overhanging the tympanic cavity, and sending off a short process, the base of which is defined in one cranium by a transverse ridge in front of the anterior articular cup for the tympanic bone. A similar process is developed in *Didunculus*, not in *Goura*, where it is barely indicated.

The presphenoid is compressed, but thickened and rounded below, where the pala-lines and pterygoids at their junction with each other abut against it: the pterygoid sends off a short process from the middle of its hinder border; but this is not met by a corresponding "pterygoid process" of the basisphenoid as in *Didunculus*.

The frontals are broad and convex, rising abruptly (as in *Didunculus*) above the coalesced cranial ends of the nasals and premaxillary (Pl. III.); in *Didus* the breadth greatly exceeds the length of the interorbital frontal convexity, as compared with *Didunculus*, and the convexity reigns in the transverse as well as the antero-posterior direction; in *Didunculus*, however, it is less concave transversely than in *Goura*. In the breadth or thickness of the interorbital septum *Didus* resembles *Apteryx* and *Palapteryx* and shows the same pneumatic cancellous structure. The posterior olfactory chambers are partially divided, as in *Dinornis*, by an upper median septum; each compartment, which is 7 lines across and an inch in length, is perforated posteriorly by an olfactory foramen more than a line in diameter, from which grooved impressions of ramifications of the nerve diverge upon the hind and upper wall of the chamber: external to the olfactory foramen is a longer one for the passage of a vein into the fore and inner part of the orbit.

The cranial ends of the nasals and nasal process of the premaxillary (Pl. XI. fig. 1, 22) are flat, depressed, thin plates; the latter at its junction with the frontal is 6 lines broad, partially divided by a median groove above and a ridge below, and by short linear fissures from the nasals: the forward extension of these bones is feebly indicated by linear grooves terminating at the outer margins of the nasal branch of the premaxillary, about 4 inches from its vertical end. The proportion of the base of the upper mandible attached to the frontal contributed by the nasals is the same as that indicated in the 'Proc. Zool. Soc.' *l. c.* The nasal branch of the premaxillary presents a full elliptical transverse section where it quits the maxillary processes, losing both depth and breadth as it recedes to join the nasals; here it retains its breadth, viz. 6 lines, but continues to be thinned off vertically to the plate above named joining the frontal. The under surface of the narrower part of the stem is angular, the upper one being gently convex.

"Where the nasal and maxillary processes diverge, there is a deep groove externally, terminating in a canal directed forwards into the rostral part or body of the premaxillary". This part is subdecurved, pointed, roughened by irregular vascular perforations and grooves, with a sharp alveolar border, which describes a sigmoid curve lengthwise, and with a deeper concavity of the palatal surface than in *Dinornis* or *Didunculus*. Moreover the concavity is partially divided lengthwise by a median ridge. The palatal surfaces of the maxillary processes and maxillaries are narrow and very convex transversely, intercepting a long narrow palato-nasal fissure. The outer side of the maxillary process is deep vertically and slightly concave lengthwise—a structure not known in *Didunculus* or any Dove, and related, like most other deviations from the Columbine cranial characteristics,

* Proc. Zool. Soc. *l. c.* p. 5.

to the provision of unwonted strength of beak in the Dodo. The maxillary branches **of** the premaxillary have completely coalesced with the maxillaries, as these have with **the** palatines; and the halves of the upper mandible here swell out laterally and more so vertically, the maxillaries rising to combine with the outer divisions of the nasals, and sending back a short process from their lower and lateral part to join the malar. The inner surface of the maxillary process (Pl. XI. fig. 1, 2⁴) is smooth and slightly convex vertically: both upper and lower borders are obtuse and thick.

The palatines arch outward from their posterior attachments, are broad and smooth mesially; the margin here is angular, with a slightly produced obtuse apex, divided by **a channel on the under surface of the palatine from the outer convex border; the** upper **and outer ridge extends forward to the maxillary; the inner** one subsides before **reaching that bone.** "The palatines form the posterior boundaries of the naso-**palatine aperture, and** approximate each other at both ends, but more closely posteriorly, yet here without meeting; whilst in *Didunculus* they coalesce before receiving the abutment of the pterygoids.

"The tympanic bone is subquadrate, with the four angles produced, and **the upper** and hinder are bifurcate, forming the double condyle for **the mastoid articulation**"[1]. There is a larger pneumatic foramen, communicating with the tympanic cavity, between the articulating cavities for these condyles.

The brain is singularly small in the present species of *Didus*; and if it be viewed as an index of intelligence of the bird, the latter may well be termed *ineptus*. The length **of the cranial cavity (Pl. XI. fig. 1, s s)** is 1 inch 8 lines, its extreme breadth 1 inch 6 lines, its greatest height 1 inch (and this is at the cerebellar fossa). The most remarkable feature in the cranial structure of *Didus* is the disproportionate size of the braincase to the important part of the neural axis it contained and protected: some approximation to this condition is made by *Dinornis*[2], the Owls, and a few large Cockatoos, e.g. *Microglossus aterrimus*; but it is fully paralleled only by the Elephant among air-breathing vertebrates, as may be seen by comparing the section Pl. XI. fig. 1 with the figures of a similar section quoted below[3].

Not only was the brain of very small proportional size in **the** present large extinct **bird,** but the division of the cranial cavity appropriate to the cerebrum proper is less in proportion to that for the cerebellum and optic lobes, at least in vertical and longitudinal diameters, than in any other known bird.

In the Elephant the thickness of the pneumatic diploë between the fore part of the cerebral cavity and that of the outer cranial wall equals the longitudinal diameter of the cavity containing the cerebral hemispheres: in *Didus* it exceeds that diameter. The thickness of the pneumatic diploë above the cerebral cavity equals the vertical diameter of

[1] Proc. Zool. Soc. 1, i. p. 6.
[2] Zool. Trans. vol. iv. pl. 34. fig. 4.
[3] Odontography, pl. 149. fig. 1; Anat. of Vertebrates, vol. ii. p. 439, fig. 296.

that cavity in *Didus*: the diploë gradually decreases in thickness as it approaches the foramen magnum. The disposition of the osseous lamellæ forming the cells or cavities of the diploë is very different in the Elephant and Dodo; they extend for the most part vertically between the outer and inner tables of the skull in the proboscidian mammal, leaving long and narrow interspaces; in the heavy ground-bird they form a congeries of small subequal and subspherical air-cells, and this structure obtains in the basal and lateral walls as well as in the superior or "roofing" **wall of the cranial cavity.** The extent of this cancellous structure at the sides of the cranial cavity may be known by the ratio of the breadth of that cavity to the breadth of the cranium, which is 3 inches **and** 8 lines at the broadest part of the brain, viz. the prosencephalon. It would seem, at first sight, as if the poorly developed brain of the Dodo had needed, on some account, unusual protection; but the true explanation rests on the size, weight, and power of the bill, and the concomitant necessity for adequate extent of attachment of the facial to the cranial part of the skull, and of the muscles from the trunk destined to sustain and wield the long and heavy-beaked head. The cerebrum of the Dodo does not greatly, and by no means proportionally, exceed the size of that part of the brain in the Crown-pigeons (*Goura*). If the great Ground-dove of the Mauritius gradually gained bulk in the long course of successive generations in that uninhabited thickly-wooded island, and, exempt from the attacks of any enemy, with food enough scattered over the ground, ceased to exert the wings to raise the heavy trunk, then, on Lamarck's principle, the disused members would atrophy, while the hind limbs, through the increased exercise by habitual motion on land, with increasing weight to support, would hypertrophy.

In the long course of generations subject to this slow rate of change, there would be nothing in the contemporaneous condition of the Mauritian fauna to alarm or in any way to put the Dodo to its wits; being, like other Pigeons, monogamous, the excitement, even, of a seasonal or prenuptial combat, might, as in them, be wanting: we may well suppose the bird to go on feeding and breeding in a lazy, stupid fashion, without call or stimulus to any growth of cerebrum proportionate to the gradually accruing increment of the bulk of the body. Whatever part of the brain was concerned in regulating or controlling muscular actions, might, indeed, be expected to show some concurrent **rate** of increase with the growing mass of the voluntary contractile fibres; and the size of the cerebellar division, (Pl. XI. fig. 1, *x o*) of the cranial cavity accords with the generally accepted physiology of the superincumbent mass of the epencephalon. The lateral depression at the fore and under part of the side of the postcerebral division of the cranial cavity indicates that the optic lobes, like the eyes, remained almost stationary during the progressive acquisition of the bulk that distinguishes the Dodo **from the** largest existing Doves.

The proportions of *Didus*, *Pezophaps*, *Casuarius*, *Rhea*, *Dromaius*, *Struthio*, *Aptornis*, *Cnemiornis*, *Palapteryx*, *Æpyornis*, *Dinornis*, &c. among terrestrial birds, of *Notornis* among the lake-haunting Coots, and of *Aptenodytes* **and** *Alca impennis* among seabirds, point to the disuse of wings in flight as the main condition of increase of size in species of birds—the next condition being absence of lethal enemies during the years requisite for such course and rate of growth.

Let foes arise from whom a power of flight is the main condition of escape, and the wingless giants of the feathered class soon succumb. Among the genera above-cited, *Apternis, Cnemiornis, Æpyornis, Palapteryx, Dinornis, Didus,* and *Pezophaps,* with the largest of the Auks, have thus passed away, while *Notornis* and *Apteryx* are on the verge of extinction through the rapid increase of population in the small island to which they are restricted. In sparsely peopled continents, such as Africa, South America, and Australia, brevipennate giants may still range the deserts, pampas, and unfrequented wilds. The ascertained recent advent of Man in New Zealand, New Britain, Ceram, Banda, Salwattie, Mauritius, Rodriguez, significantly points to the conditions under which have come to pass, in lapse of time, so strange an anomaly as a bird with the specially modified instruments of flight reduced below the power of exerting that mode of locomotion, yet, as a bird, retaining the conditions of the respiratory and tegumentary systems of the volant class, of which it has become a degenerate member. With the cessation of the chief of those conditions, viz. the absence of enemies, such birds necessarily perish.

Refraining, however, from further indulgence in an easy and seductive vein of speculation, I would recall attention to the notable protuberance in the cranial cavity of the Dodo (Pl. XI. fig. 1, *o*) developed towards the upper part of the vertical tentorium, contracting at its lower part into the ridge dividing the prosencephalic from the mesencephalic chamber. In the latter are the orifices for the issue of the trigeminal nerve, the larger and posterior (ib. *tr*) giving passage to the third and second divisions, and answering to the combined foramen ovale and rotundum of mammals, and the smaller and anterior foramen dismissing the first or orbital division of the fifth nerve. At the upper part of the mesencephalic fossa the narrow groove for the lateral venous sinus impresses and defines the back part of the tentorial protuberance, above which it bifurcates, the lower branch bounding or defining the wall of the superior semicircular canal and the upper part of the primitive acoustic capsule. Below this arch is an oblong cerebellar fossa (ib. *n*) which appears to have received veins from the cranial diploë. Beneath this fossa, and just behind the mesencephalic chamber, is the multiperforate internal auditory depression. Next behind this is the outlet for the vagal nerve and entojugular vein. Below this are the small precondyloid foramina. There is a falcial ridge, low and thick, indicating the division of the prosencephalic chamber into lateral compartments for hemispheres ; and this ridge shows a narrow groove as for a small longitudinal sinus. A transverse linear groove abruptly defines the fore part of the ridge.

The vertically expanded anterior part of the premaxillary (ib. fig. 1, *x*) has a large pneumatic cavity communicating by a reticulate wall with the cells of a cancellous structure, larger than those of the cranial diploë. The maxillary branch of the premaxillary (ib. 22*) consists of a light open-work air-diploë, with a very thin outer case of bone. The short symphysis mandibulæ shows a small cavity, surrounded by more minutely cancellous structure and thicker compact walls, especially at the upper and hinder parts.

Although some characters have been too much insisted on (e. g. the "superoccipital foramen") as exemplifying the affinity of the Dodo, the more essential characters of the skull relate to its true Columbine character, while the deviations from that part of the skeleton of volant Doves are explicable in the adaptive developments needed for the wielding of long, powerful, massive mandibles, serving most probably to enable the bird to subsist on some proportion of animal diet, in addition to such vegetable food as it might gain from the ground. Such indiscriminate feeding doubtless rendered its flesh less palatable than that of the winged Pigeons of the Mauritius to the Dutch navigators of the sixteenth and seventeenth centuries.

But the affinities of *Didus* will be more fully and decisively brought out in the comparison of the, in this respect, more instructive and light-giving parts of the skeleton.

§ 3. *Comparison of the Skeleton.*

The dorsal region of the vertebral column shows, in some birds, a confluence of certain vertebræ: I have observed four to be so welded together by both centrums and neural spines in *Phœnicopterus*, viz. the second to the fifth dorsal inclusive, leaving the sixth free, which articulates with the first costigerous sacral vertebra. In *Platalea* three dorsals coalesce in advance of the antepenultimate free vertebra. In the smaller diurnal birds of prey five dorsal vertebræ are usually confluent, leaving one free vertebra for the lateral movements of the trunk between such dorsal "sacrum" and the pelvic one. In Vultures, Plovers, Bustards, Cranes, *Psophia*, *Cariama*, *Palamedea*, Auks, Penguins, and in all flightless land-birds save the Dodo, no such anchylosis takes place. The *Columbidæ* are the species in which the dorsal vertebræ, homologous and the same in number with those of *Didus*, undergo the process of confluence into one mass of bone: they are the three which immediately precede the last (moveable) dorsal vertebra; and of these the two anterior develope, in *Goura* and *Didunculus*, hypapophyses closely corresponding in shape and proportion with those in the Dodo.

The chief difference which *Didus* offers in the present region of the vertebral column from that of *Columbidæ* is in the greater number of the vertebræ or segments which are typically completed by bony hæmapophyses articulating with pleurapophyses and directly with their mass of coalesced and expanded hæmal spines constituting the sternum. Of these typical thoracic segments there were five in *Didus* (Pl. III.); *Didunculus* (ib.) shows four; *Goura* three. In both existing genera these segments are succeeded by a single one, anchylosed to the fore part of the sacrum, but with the pleurapophysis long and moveable, with its hæmapophysis terminating in a point before reaching the sternum, and extensively connected with the antecedent hæmapophysis or sternal rib: in both genera two dorsal vertebræ in advance of the typically complete one

6

have moveable pleurapophyses terminating freely in a point, with no hæmapophyses other than the costal processes of the sternum may represent. In *Goura*, which has six pairs of moveable or thoracic ribs, the second pair belong to the first of the three anchylosed dorsal vertebræ: in *Didunculus*, which has seven pairs of thoracic ribs, the second pair belongs to the free dorsal immediately in advance of the anchylosed mass. Supposing *Didus* to have had one pair of ribs behind, and two pairs in front of those that directly articulate with the sternum, as the vertebra Pl. V. fig. 7 indicates, it would would have had eight pairs of thoracic ribs; and I think this excess of one pair beyond the formula in *Didunculus* to be very probable in the large-bodied, small-winged, extinct Ground-dove.

As far as the series of Dodo's neck-vertebræ under my observation exhibit such characters, the proportion of those with neural spines, or with hypapophyses, or both, is the same as in the *Columbidæ*. In this family, as in most birds, the greater part of the series want both processes. The cervical parapophyses, descending to form the sides of the carotid canal, do not meet, coalesce, and circumscribe it in any cervical vertebra of *Goura* or *Didunculus*; and not any of the vertebræ of *Didus*, which I have yet received, shows such circumscription of the hæmal canal. The majority of the cervicals in *Didus* (those, viz., that lack both neural spines and hypapophyses) are broader and more massive in proportion to their length than in the winged Doves. The third cervical in *Didus* has both the above processes, as in *Columbidæ*: the characters of the axis vertebra in the same family are closely repeated in that of the Dodo. In the Raptores the axis vertebra is shorter in proportion to its length, and a greater proportion of the cervical vertebræ at both ends of the series have both neural spines and hypapophyses.

The ribs of the Dodo are as broad, in proportion to their length, as in Doves, but are relatively longer in proportion to the dorsal region, encompassing a more capacious thoracic-abdominal cavity. The ribs of the Vulture are more expanded than in *Didus*, especially where they afford the extensive attachment to the epipleurals. But I shall not dwell further on the comparative characters of this part of the skeleton, as more decisive ones of the affinity of *Didus* are afforded by other parts.

In comparing the sternum of the Dodo with that of Doves of flight, the first well-marked difference is in the adaptive development of the keel in the last (Pl. III. fig. 2, *Didunculus*), and in the provision for the concomitantly broader coracoids, the grooves for which meet and run into each other across the fore part of the bone in existing *Columbidæ* (Pl. XII. fig. 2, *b*); consequently the inner or upper wall of the confluent grooves forms a median prominence (ib. *e*) at the front margin of the sternum, contrasting with the wide notch at that part of the bone in the Dodo (Pl. IV. fig. 4). The next difference, as compared with *Goura* and most Pigeons, is the absence of the entolateral processes (Pl. XII. fig. 2, *i*) in the Dodo's sternum: but *Didunculus* singularly exemplifies its nearer affinity to *Didus* by a like absence of those processes; only the sternal margins behind the ectolateral processes (ib. fig. 1, *h*), instead of converging with a slight convexity to an obtuse apex, as in Pl. VI., describe a concavity, through an expansion of the posterior truncate end of the breast-bone. The sternum of *Didunculus* may be said to show one pair of posterior notches (Pl. XII. fig. 1, *f*), that of other Pigeons two pairs (ib. fig. 3, *f f*); but the sternum of *Didus*, which is relatively broader, shows no other trace of the anterior notch (Pl. VI. *f*) than is afforded by the rounded angle at which the ectolateral process (*h*) rises from the bone. Although the costal margin is relatively shorter in Doves of flight than in the Dodo, again an intermediate condition is manifested by *Didunculus* as compared with *Goura*, in which latter Dove there are articular surfaces for three sternal ribs (Pl. XII. fig. 3, *c* 1, 2, 3), whilst in *Didunculus* there are four (ib. fig. 1, *c*). *Didunculus* also exhibits, more strongly than *Goura*, the obtuse ridges (ib. fig. 2, *r*) converging like buttresses from the outer wall of the coracoid groove to the fore part of the keel, where they subside. In *Didunculus* there is a pneumatic foramen exterior to the coracoid groove, corresponding with *p*, fig. 4, Pl. IV., which I do not find in the sternum of *Goura*; but in the Crown-pigeons the pneumatic foramina along the middle line of the upper surface of the sternum are conspicuous; they are confined to the fore part of that surface in *Didunculus* (Pl. XII. fig. 1).

In the direction of the ectolateral processes *Goura* (ib. fig. 3, *h*) is intermediate between *Didunculus* and *Didus*. The pectoral ridge on the outer surface of the sternum, continued backward from the outer end of the coracoid groove, is adaptively better marked in Pigeons of flight than in the Dodo; and the pair of ridges are more nearly parallel in their backward course, not so convergent as in *Didus*. In *Goura* the subcostal ridge is better marked than in *Didunculus*. In no Dove of flight is the body of the sternum so broad and hollow as in *Didus* (Pl. XI. fig. 4); in this respect the Vulture more nearly resembles the Dodo, as it does also in the more convex anterior contour of the keel: but the vulturine sternum does not lose breadth as it extends backward; it is a square-shaped shield in birds of prey, shorter in proportion to its breadth, with a greater extent of costal process and margin, and with the ectolateral processes, when they exist, extending backward as far as the hinder border of the bone. In the thorough quest of resemblances to the Dodo's sternum which I have made through the class of Birds, I came upon an unexpected superficial likeness to it in the sternum of a Night-jar (*Podargus humeralis*). The ectolateral processes (Pl. XII. fig. 4, *h*) rise behind the moderately extended costal borders, *c*; and beyond them the body of the sternum converges to an obtuse end, with a contour similar to that in *Didus*. Moreover the coracoid grooves are divided from each other by a free concave border, less deep and extensive, indeed, than in *Didus*, but as free from any trace of episternal projection. The ectolateral processes, however, are extended backward to beyond the sternal body; and this part usually shows a pair of small ectolateral notches, *f*, of which one was present on one side in the specimen figured.

Through the reduction of the coracoids in all flightless birds, there is an interval between their sternal articulations: this is long and concave in the Dodo, but is longest and most deeply concave in *Apteryx*; it is long but almost straight in *Rhea*; in *Casuarius* and *Dromaius* it is narrow but deeply notched; in *Struthio* it developes a short episternal process. In no Grallatorial sternum with both ecto- and ento-lateral processes (as e.g. *Otis*, *Œdicnemus*, *Charadrius*) do the former project, as in *Didus* and the Rasores, immediately behind the costal margin, but they are continued, parallel with the keel, from the outer and posterior angle of the sternum, distant from the costal margin. In old Plovers the entolateral process joins the contiguous angle of the sternal body, and converts the inner notch into a foramen.

In the breast-bone of the Dodo we plainly discern the Columbine modification of the Gallinaceous type, simplified in the minor development of those parts relating adaptively to the power of flight, and expanded and excavated for the support of the larger gizzard with its heavier grindstones[1].

In comparing the pelvis of *Didunculus* and *Goura* (Pl. XII. fig. 5) with that of *Didus* (Pl. VII. fig. 1), the correspondences are:—in the general shape, proportions and disposition of the ilia; in the articulation therewith of the last pair of moveable ribs, and of the short straight confluent pleurapophyses of the three succeeding sacral vertebræ; then follow, as in *Didus*, three vertebræ without pleurapophyses, these reappearing in the next two with their extremities converging to abut against a prominence of the inner surface of the ilium in the same relative position. The difference here is in the two equal and more slender rib-buttresses, in place of the single stronger one, which is the more common structure in *Didus*; but in *Goura* I have noted an instance in which it agreed with the *Didunculus* on the left side, and with *Didus* on the right, in the last-specified character. In the Crown-pigeons, also, there is an indication of the transverse ridge marking off the under part of the centrum of the first sacral from the rest, and those that follow are less expanded than in the Dodless; moreover in *Didunculus* they show a median canal instead of a ridge, while the ridge is feebly indicated here and there and there is no canal in *Goura*. In neither *Didunculus* nor *Goura* do the sacral centrums behind the last rib-abutments diminish in breadth so suddenly as in *Didus*: in both the winged Pigeons the hinder part of the pelvic cavity is relatively deeper and narrower than in *Didus*; in both, also, the upper and anterior concave tracks of the ilia are deeper; and in *Didunculus* the mesial borders do not attain the neural crest, but leave a pair of open longitudinal canals at that part of the pelvis; in *Goura* those margins reach the neural crest, but do not overtop it at any part. In *Goura* the acetabula are more in advance of a median position than in *Didunculus*, *Columba magnifica*, or *Didus*. Although the ischiadic foramina are completed by terminal confluence of the ilium and ischium in

[1] The habit of the Dodo to avail itself of extraneous crushers to a gallinaceous or struthious degree, is attested by the quotation, p. 8, not the least interesting of the fruits of the extensive research of the learned and conscientious author of the Article Dodo, in the 'Penny Cyclopædia.'

Dromaius and *Casuarius*, yet the length of those foramina (which are unclosed) in *Struthio* and *Apteryx*, concomitant with the greater relative length of the pelvis, shows the difference of *Didus* from the cursorial Brevipennates in this part of the skeleton. The ischia of the winged Pigeons resemble those of the Dodo; but the inner longitudinal ridge is more strongly marked in *Didunculus*: in the *Goura* it is less developed than in *Didus*; the bone is longer also in proportion to its breadth, and the ischiadic foramen is longer and narrower: the proportions of that in *Didunculus* are more like those in *Didus*. In *Didunculus* the pubis coalesces with the ischium behind the small obturator foramen, but leaves a second or posterior elongate ischio-pubic vacuity. The greatest amount of resemblances with the pelvis of the Dodo is found in that of different members of the Dove-tribe.

In comparing the pelvis of the Dodo with that of the Vulture (Pl. XII. fig. 6), we find in the latter that the first two confluent sacral vertebræ supporting moveable ribs are succeeded by several with short abutting ribs, the extent of this part of the sacrum being nearly one-half of the whole, instead of one-fourth as in *Didus* and the Doves. The reappearance of rib-abutments after four ribless sacrals is in the posterior third of the sacrum, and they are continued to the end of that bone from the last four vertebræ of the series, constituting a very marked difference, both as to number and the character of the vertebræ in the sacral part of the pelvis.

With regard to the iliac bones, the anterior concave track occupies two-thirds of the extent of the bone in *Vultur*, not one-half as in *Didus* and most Doves; the breadth of the posterior parts of the ilia with the intervening sacrum in the Vulture is relatively less than in the winged Doves, and differs in a greater degree from that characteristic part in the sacrum of *Didus*. In *Ciconia* the antacetabular part of the pelvis is relatively longer, and the iliac bones are more expanded anteriorly. In *Platalea* the proportions are more nearly those in *Didus*. In *Otis* the ilia touch the fore part of the sacro-spinal ridge, but leave both posterior and anterior apertures of the ilio-neural canals widely open. In *Œdicnemus* and *Charadrius* they are grooves, the ilia not reaching the sacral spines. The external concavity of the ilium is longer, narrower, and deeper, in most waders, than in *Didus*. In *Eudyptes* and *Aptenodytes* the ilia are more expanded anteriorly, but the whole pelvis is narrower and longer than in *Didus*. The Gar-fowl (*Alca impennis*)[1], *Uria*, *Podiceps*, and *Colymbus*, all show still longer and narrower proportions of the pelvis.

In the Doves of flight the proportions and relative position of the three compartments of the cranial cavity differ from those in the Dodo. Both the pros- and mes-encephalic ones are proportionally larger than the epencephalic; and the mesencephalic compartment lies more directly below the prosencephalic one. A very thin stratum of finely cellular diploë divides the two tables of the skull along the medial line of the upper surface: it is thicker between the orbits. The falcial ridge at the inner surface

[1] Trans. Zool. Soc. vol. v. pl. 51.

of the prosencephalic roof resembles that in *Didus*. The tentorial ridge bifurcates half-way down, the front portion dividing, almost horizontally, the pros- from the mesen-cephalic compartment, the hinder and more obtuse ridge dividing, almost vertically, the mes- from the epencephalic compartment. The angle of bifurcation is slightly produced and obtuse, but represents very feebly the tentorial tuberosity (Pl. XI. fig. 1, *o*) in the Dodo: from it, in *Goura*, is continued backward the arch of bone formed by the superior semicircular canal, above which is the groove for the venous sinus, as in *Didus*. The internal auditory fossa is less deep than in *Didus*: above it is a similarly vertically oblong cerebellar pit. The nerve-foramina correspond with those in *Didus*: the ento-carotid canal opens into a rather deeper sella in *Columba palumbus*.

On comparing the cranial cavity, as exposed by a vertical longitudinal section in the Dodo (Pl. XI. fig. 1), with that of a Dinornis similarly exposed[1], the first difference is the smaller proportional depth of the diploë in the larger wingless bird, which is not greater over the prosencephalic than over the epencephalic compartment; next may be noticed the larger relative size of the former compartment, indicating the larger cere-brum of the Dinornis, then the absence of the tentorial tuberosity, the sharper and more produced superior part of the tentorial ridge arching transversely between the cerebrum and cerebellum, the smaller internal auditory fossa, and the deeper sella: the mesencephalic compartment, or cavity for the optic lobe, is less in proportion to the prosencephalic compartment than in *Didus*; it holds, however, a similar relative posi-tion: finally, the cerebellar pit, above the internal auditory fossa, is wanting in the Dinornis.

The Dodo agrees with the Doves in possessing a slender furculum, forming an acute angle: it resembles *Columba galeata*, more especially, in the halves of that bone being united by ligament below, and forming separate styles or "clavicles."

The humerus of the Goura closely repeats most of the characters described in that of the Dodo: but its length is proportionally greater, being 3 inches 9 lines, nearly equal to that of the sternum or pelvis, whereas the humerus of the Dodo is little more than half the length of either sternum or pelvis. The processes for the attachment of the muscles are, nevertheless, fully as strongly developed in *Didus* (Pl. VIII. figs. 12 & 14) as in the volant Doves (Pl. XII. figs. 8 & 9, *Goura*); that, indeed, which is a ridge (*r*) on the back part of the shaft in *Didus*, is a mere rough surface in *Goura*, and does not show in *Didunculus*. The pneumatic fossa, which varies in depth in the two humeri of the Dodo, is in both relatively larger and shallower than in *Goura*. The pectoral process is thinner, but relatively rather more produced, in *Didunculus*. The humerus in *Œdi-cnemus*, *Otis*, and *Charadrius* has a more longitudinally extended, thinner, and more produced pectoral ridge than in *Didus* and the *Columbidæ*; there is a more marked ectocondyloid tuberosity, which in *Charadrius* becomes a pointed process.

There is nothing to be gained by giving the details of the more striking differences

[1] Trans. Zool. Soc. vol. iv. pl. 24, fig. 1.

which the humerus presents in Penguins, Auks, and birds of prey, as compared with that bone in the Dodo; but a few words may be recorded of the comparison of the humerus of the Dodo with that of the flightless bird of New Zealand so nearly approaching to it in size, which bird is described in the 5th volume of the 'Transactions' of the Society under the name of *Cnemiornis* (p. 395, pl. 66. figs. 7–10). In that extinct species, although the humerus is 5½ inches in length, the parts indicative of the forces by which it was worked are comparatively feebly developed. The ulnar tuberosity is narrower, thicker, more obtuse, and its base has neither the upper nor lower excavation: it rises above the articular head, which is less prominent and narrower than in *Didus*; the pectoral ridge is shorter and situated lower down upon the shaft, not on the same level with the radial tuberosity as it is in *Didus*; the distal articulation is of the same size as in *Didus*, but neither the radial nor the ulnar convexity is so prominent or well-defined.

The ulna of the Dodo is shorter absolutely, and much more so proportionally, than in the Goura and most other volant Doves. In these it exceeds the humerus by about one-fourth its own length; in *Didunculus* (Pl. III.) it is a little longer than the humerus; in the Dodo (ib.) it is shorter than the humerus. The length of the ulna in *Goura coronata* is 4 inches 6 lines; it is more bent than in the Dodo; the quill-tubercles, seven or eight in number, are more prominent; nevertheless the rough depression for the insertion of the chief flexor is less deep and less defined. The plumed winglet of the Dodo would seem, therefore, to have been frequently and forcibly moved.

In comparing the femur of the Dodo with that of the largest Dove, the bone appears gigantic. The length of the femur in *Goura coronata* (Pl. XII. fig. 11) is but 3 inches 3 lines, and it is more slender in proportion to its length than in the Dodo; it, however, repeats the few characteristics, if they may be so termed, of the Dodo's femur. It has the pneumatic foramen in the same position, perhaps proportionally larger; it has the same large oblong surface for the ligament at the head of the bone; the great trochanter has the same form and disposition, but is not quite so much produced anteriorly; there is a slight depression instead of a ridge for the trochanter minor; the fore part of the inner condyle is relatively thicker and less produced. The femur in *Otis* and *Œdicnemus* has a thicker and shorter trochanter major, a more narrow and shallow rotular channel; it is shorter in comparison with the tibia, and more especially with the metatarsus, than in *Didus* and the Doves.

The femur of *Aptornis otidiformis*[1] is of the same size as that of the Dodo; but it has no pneumatic foramen, the head is more hemispheroid and inclined forward, the ligamentous pit is deeper and more circular, the supracervical articular surface is not defined from that of the head, there is a wider and deeper depression at the fore part of the proximal end of the femur, and a more prominent tuberosity on the back part; the ridge continued from the back part of the shaft to that of the inner con-

[1] Trans. Zool. Soc. vol. v. pl. 65. fig. 3.

dyle is more produced and sharper in *Aptornis*, the fore part of the same condyle is less produced.

The femur in *Cnemiornis*[1] and *Dinornis*[2] is much thicker, in proportion to its length, than in either *Aptornis* or *Didus*. In *Pezophaps* the great trochanterian ridge rises higher above the neck, and the shaft has a more uniform thickness, with the inner contour less concave, than in *Didus*.

The characters which have been noted at the proximal and distal ends of the tibia of *Didus* are repeated in those of the tibia of the Goura. The difference in size is more marked than in the femur; the length of the tibia of *Goura coronata* is 4 inches 7 lines, and its shaft is more slender, in proportion to its length (Pl. XII. fig. 13), than in *Didus* (Pl. X.). The tendency to a trihedral form of the shaft is less marked in *Goura*; the anterior prominences of the distal condyles are thicker in proportion to the intervening fossa.

In the Vulture the fibular ridge is more parallel with the long axis of the shaft than in *Didus*; the tendinal canal is less cylindrical, has an oblique course from the middle of the anterior surface towards the inner condyle; the fore parts of both distal condyles are less produced and less convex; the distal end is narrower from before backwards in proportion to its breadth; both extremities of the bone are less expanded in proportion to the shaft than in the Dodo.

In the great Plover (*Œdicnemus crepitans*) the tibia, as in other Grallæ, is longer in proportion to its thickness than in *Didus*; the epicnemial process rises higher above and projects further in front of the condylar surfaces before it divides into the pro- and ectocnemial plates; and these are relatively more produced. The fibular ridge is shorter in proportion to the length of the tibia, is more prominent, and more parallel with the axis of the shaft. The distal condyles project further backward than in *Didus*. The tibia in *Charadrius, Otis, Tantalus, Grus, Ciconia, Mycteria, Porphyrio,* opposes similar or equivalent differences to those in *Œdicnemus*, against the affinity of *Didus* to any of those Grallæ.

In the comparison of the tibia of this extinct flightless bird with that of the *Cnemiornis*, the wonderful development of the plates and processes at the proximal end of the bones in the New Zealand bird is strikingly manifested. In *Cnemiornis* the fibular ridge runs in a line with the shaft, and does not incline from above obliquely forward as in *Didus* and the Doves; the ridge on the outer side of the distal fourth of the bone is stronger and sharper in *Cnemiornis*; the tendinal canal is transversely elliptical, medial in position, with a slight inward inclination; the intercondyloid fossa is much wider in *Cnemiornis*. The differences, indeed, in all the characters of the tibia, as compared with *Didus*, in the Vultures, Plovers, Penguins, and terrestrial flightless birds tend to render more instructive and convincing the resemblances which Pigeons present in the same characters to the extinct Mauritian bird.

[1] Trans. Zool. Soc. vol. v. pl. 65. fig. 1.   [2] Ibid. fig. 6.

§ 4. *Conclusion.*

The affinities or place in nature of the Dodo being thus determined by the characters of its skeleton, but few words remain to be said on the bearings of present knowledge of this species upon other zoological generalizations.

The researches and observations of naturalists have been carried out to such an extent as to support the conclusion that the *Didus ineptus* does not now live in any part of the world, and that it never existed save in that part of which the island of Mauritius may be a remnant. Consequently the species there originated; and the most intelligible conception of its mode of origin is that to which I have alluded in the description of the brain-case (p. 39).

The Dodo exemplifies Buffon's idea[1] of the origin of species through departure from a more perfect original type by degeneration; and the known consequences of the disuse of one locomotive organ and extra use of another indicate the nature of the secondary causes that may have operated in the creation of this species of bird, agreeably with Lamarck's philosophical conception of the influence of such physiological conditions of atrophy and hypertrophy[2]. The young of all Doves are hatched with wings as small as in the Dodo: that species retained the immature character. The main condition making possible the production and continuance of such a species in the island of Mauritius was the absence of any animal that could kill a great bird incapable of flight. The introduction of such a destroyer became fatal to the species which had lost such means of escape[3]. The Mauritian Doves (*Columba nitidissima* and *C. meyeri*) that retained their powers of flight continue to exist there.

As I have no reason to offer why one kind of Pigeon should have retained and another lost its powers of flight, nor am I able to adduce a particle of evidence of the hypothetical degrees of diminution of the wing-bones to their stunted proportions in *Didus*, any more than in *Dinornis*, I feel that in the foregoing remarks I lay myself open to the rebuke of fellow-labourers who may think with the able authors who last treated of the present subject.

They warn their readers to " beware of attributing anything like *imperfection* to these anomalous organisms, however deficient they may be in those complicated structures which we so much admire in other creatures. Each animal and plant has received its peculiar organization for the purpose, not of exciting the admiration of other beings, but of sustaining its own existence. Its perfection, therefore, consists, not in the number or complication of its organs, but in the adaptation of its whole structure to the external circumstances in which it is destined to live. And, in this point of view, we shall find that every department of the organic creation is equally perfect, the

[1] Histoire Naturelle, &c., 4to, tom. xiv. " Dégénération des Animaux :" 1760.

[2] Philosophie Zoologique, 8vo, 1809, tom. i. chaps. 3, 6, & 7.

[3] Agreeably with the principle of the " contest for existence " by which I explained the extinction of the species of *Dinornis*, Trans. Zool. Soc. vol. iv. p. 14, 1851.

H

humblest animalcule or the simplest conferva being as completely organized with
reference to its appropriate habitat and its destined functions as Man himself, who
claims to be lord of all. Such a view of the creation is surely more philosophical than
the crude and profane ideas entertained by Buffon and his disciples"[1].

Nevertheless the truth, as we have or feel it, should be told. In the end it may prove
to be the more acceptable service. The *Didus ineptus*, L., through its degenerate or
imperfect structure, howsoever acquired, has perished. What have the stigmatizers of
Buffon to offer in lieu of his theory as applied to the origin of this species of bird?
They begin by asking, "Why does the whale possess the germs of teeth which are never
used for mastication? and why was the Dodo endowed with wings at all, when those
wings were useless for locomotion? This question," they own, "is too wide and too deep
to plunge into at present." They nevertheless proceed to remark, "These apparently
anomalous facts are really the indications of laws which the Creator has been pleased
to follow in the construction of organized beings; they are inscriptions in an unknown
hieroglyphic, which we are quite sure mean *something*, but of which we have scarcely
begun to master the alphabet. There appear, however, reasonable grounds for believing
that the Creator has assigned to each class of animals a definite type or structure, from
which He has never departed, even in the most exceptional or eccentric modifications of
form. Thus, if we suppose, for instance, that the abstract idea of a Mammal implied
the presence of teeth, and the idea of a Bird the presence of wings, we may then
comprehend why in the Whale and the Dodo these organs are merely *suppressed*, not
wholly *annihilated*"[2].

This notion of typeforms or centres, unfortunately, has not merely relation to abstract
biological speculations or theories, but to practical questions on which the true progress
of Natural History vitally depends. If such types do exist, the National Museum, it is
argued, may be restricted to their exhibition: and so our legislators and the public were
assured by the Professor of Natural History in the Government School of Mines[3], when
the question was before the "House" four years ago. I have let slip no suitable occasion[4] to combat and expose what has seemed to me to be both an erroneous and mis-
chievous view, most obstructive to the best interests of the science; and, standing alone

[1] Strickland and Melville, 'The Dodo and its Kindred,' 4to, 1848, p. 34.

[2] *Op. cit.* p. 34.

[3] See letter in 'The Times' of May 21st, 1862, advocating the limitation of the National Museum of Natural
History to "six rooms," signed THOMAS H. HUXLEY, F.R.S.

[4] Reply to the above in 'The Times' of May 2nd, 1866, and in both editions (1861, 1862) of my 'Discourse
on the Extent and Aims of a National Museum of Natural History.' "Some naturalists urge that it is only
necessary to exhibit the typeform of each genus or family. But they do not tell us what is such 'typeform.' It is a metaphysical term, which implies that the Creative Force had a guiding pattern for the construction of all the varying or divergent forms in each genus or family. The idea is devoid of proof; and those
who are loudest in advocating the restriction of exhibited specimens to 'types' have contributed least to lighten
the difficulties of the practical curator in making the selection." (Id. 1862, p. 24; see also pp. 26-31.)

as I seemed to do on this point in the array of evidence before the "Parliamentary Committee on the British Museum, 1860," I was glad to find my views on type-forms adopted and paraphrased by the President of the British Association in his Inaugural Address at the Meeting at Nottingham[1], in the present year.

DESCRIPTION OF THE PLATES.

PLATE I.

Ideal Scene in the island of Mauritius before its discovery, in 1598, by the Dutch, founded on :—

Fig. 1. Picture of the Dodo, by Roelandt Savery, 1626, in the Royal Gallery of Berlin.

Fig. 2. Fac-simile of R. Savery's Picture of the Dodo, in the possession of the late Wm. J. Broderip, Esq., F.R.S. (no date).

Fig. 3. Picture of the Dodo, by R. Savery, 1628, in the Imperial Collection of the Belvedere, Vienna.

Each figure is coloured, and of the exact size, as in the original paintings.

PLATE II.

Two views of the Dodlet (*Didunculus strigirostris*, Peale; *Gnathodon*, Jardine), natural size, from the living bird, obtained at the Samoan or Navigators' Islands, and transmitted from Sydney, New South Wales, by George Bennett, M.D., F.L.S.[2], to the Gardens of the Zoological Society of London, in 1864, where the paintings, of which the above are fac-similes, were made for the present work. A sketch of the dried head of the Dodo in the Ashmolean Museum, Oxford, of rather less than half the natural size, is introduced into the picture, now in the Author's possession[3].

[1] "The doctrine of typical nuclei seems only a mode of evading the difficulty. Experience does not give us the types of theory ; and, after all, what are those types ? It must be admitted there are none in reality. How are we led to the theory of them ? Simply by a process of abstraction from classified existences. Having grouped from natural similitudes certain natural forms into a class, we select attributes common to each member of the class, and call the assemblage of such attributes a type of the class. This process gives us an abstract idea ; and we then transfer this idea to the Creator, and make Him start with that which our own imperfect generalization has derived." (Address, &c., by WILLIAM R. GROVE, Esq., Q.C., M.A. 8vo, London, 1866 ; p. 31.)

[2] See Dr. Bennett's excellent notes on the living *Didunculus*, in the 'Proceedings of the Zoological Society of London,' 1864, p. 139.

[3] To my friend Dr. Bennett I owe the first specimens of the *Nautilus pompilius*, impregnated uterus of the Kangaroo and Ornithorhynchus, the young Ornithorhynchus, and other rare subjects of early Memoirs. Natural History owes much to this accomplished and indefatigable Observer.

PLATE III.

Fig. 1. Side view of the skeleton of the Dodo (*Didus ineptus*, L.), with an outline of the bird as represented in the oil-painting presented to the British Museum by Edwards, Naturalist and Librarian of the Royal Society, into whose possession it came at the decease, in 1753, of Sir Hans Sloane, P.R.S., with the statement, or tradition, that the painting had been made, of the natural size, from a living specimen of the Dodo, in Holland. The bones represented in profile, of the natural size[1], testify to the accuracy of the form and proportions of the Dodo given in the painting.

Fig. 2. An outline of the Samoan Dove or Dodlet (*Didunculus strigirostris*, Peale; *Gnathodon strigirostris*, Jardine[2]), of the natural size, from the specimen sent by Dr. G. Bennett, and living, in 1864, in the Gardens of the Zoological Society of London, with a view of the skeleton, corresponding with that of the Dodo.

PLATE IV.

Fig. 1. Front view of the fourth (or first of the three confluent) dorsal vertebræ (centrum and neural arch).

Fig. 2. Vertebral rib, or pleurapophysis, of the same vertebra, front view.

Fig. 3. Sternal rib, or hæmapophysis, of the same vertebra: *a*, outer side; *b*, upper or pleural end; *c*, lower or sternal end; *d*, front margin; *e*, inner surface.

Fig. 4. Front view of sternum, or connate mass of hæmal spines, including that of the same (fourth dorsal) vertebra.

Fig. 5. Inner surface of an anterior pleurapophysis, with coalesced appendage, *a*.

Fig. 6. Oblique view of ditto, ditto.

Fig. 7. Anterior pleurapophysis, with appendage, *a*, front view: *c*, capitular end; *d*, tubercular end; *f*, hæmal end; 7 *a*, outer surface; 7 *b*, inner surface.

Fig. 8. An anterior pleurapophysis, front view.

Fig. 9. Posterior surface of the upper end of a posterior pleurapophysis: 9 *a*, body and lower end of ditto.

Fig. 10. Part of a pleurapophysis which has been broken and healed.

Fig. 11. Lower end of a posterior dorsal pleurapophysis, with connate rudiment of appendage, *a*.

Fig. 12. Hæmapophysis.

[1] The scapular arch is rotated in advance of the ribs to show the character of the anterior dorsal vertebræ.

[2] See also Gould, 'Birds of Australia,' part 22 (March, 1846).

PLATE V [1].

Fig. 1. Fourth, fifth, and sixth dorsal vertebræ, anchylosed, side view.
Fig. 2. Ditto, ditto, upper view.
Fig. 3. Ditto, ditto, under view.
Fig. 4. Ditto, ditto, back view.
Fig. 5. Ditto, ditto, mutilated, of another Dodo.
Fig. 6. Anterior dorsal vertebra, side view.
Fig. 7. Ditto, front view; *pl,* outline of heads of floating rib.
Fig. 8. Penultimate cervical vertebra, side view.
Fig. 9. Ditto, back view.
Fig. 10. Middle cervical vertebra, upper view.
Fig. 11. Ditto, under view.
Fig. 12. Axis, or second cervical vertebra, upper view.
Fig. 13. Ditto, under view.

PLATE VI.

Fig. 1. Under view of sternum.
Fig. 2. Upper or inner view.
Fig. 3. Back view.

PLATE VII.

Fig. 1. Under or inner view of pelvis.
Fig. 2. Upper or outer view of pelvis.

PLATE VIII.

Fig. 1. Middle cervical vertebra, upper view.
Fig. 2. Fifth cervical vertebra, upper view.
Fig. 3. Fourth cervical vertebra, under view.
Fig. 4. Right coracoid and clavicle.
Fig. 5. Left coracoid and clavicle.
Fig. 6. Right scapula, outer view.
Fig. 7. Right scapula, inner view.
Fig. 8. Left moiety of scapular arch, outer view.
Fig. 9. Ditto, inner view.
Fig. 10. Upper articular end of right coracoid.
Fig. 11. Lower ditto.
Fig. 12. Left humerus, anconal or back surface.

[1] I beg to return my acknowledgments to the Trustees of the Liverpool Museum for the opportunity of figuring two specimens, in this Plate, from the collection of Dodos' bones in that Museum.

PLATE XII.

Fig. 1. Sternum of *Didunculus*, upper view.
Fig. 2. Ditto, front view.
Fig. 3. Sternum of *Goura*, upper view.
Fig. 4. Sternum of *Podargus humeralis*, under view.
Fig. 5. Pelvis of *Goura*, under or inner view, half natural size.
Fig. 6. Pelvis of *Gyps* (Vulture), under or inner view, half natural size.
Fig. 7. Left moiety of scapular arch, *Goura*.
Fig. 8. Left humerus of *Goura*, anconal surface.
Fig. 9. Ditto, palmar surface of upper end.
Fig. 10. Ditto, palmar surface of lower end.
Fig. 11. Right femur of *Goura*, front view.
Fig. 12. Ditto, back view of upper end, and back view of lower end.
Fig. 13. Right tibia and fibula of *Goura*, front view.

All the figures are of the natural size, save when otherwise expressed. The letters are explained in the text.

THE END.

PRINTED BY TAYLOR AND FRANCIS, RED LION COURT, FLEET STREET.

BLANK
PAGE

PLATE II

BLANK
PAGE

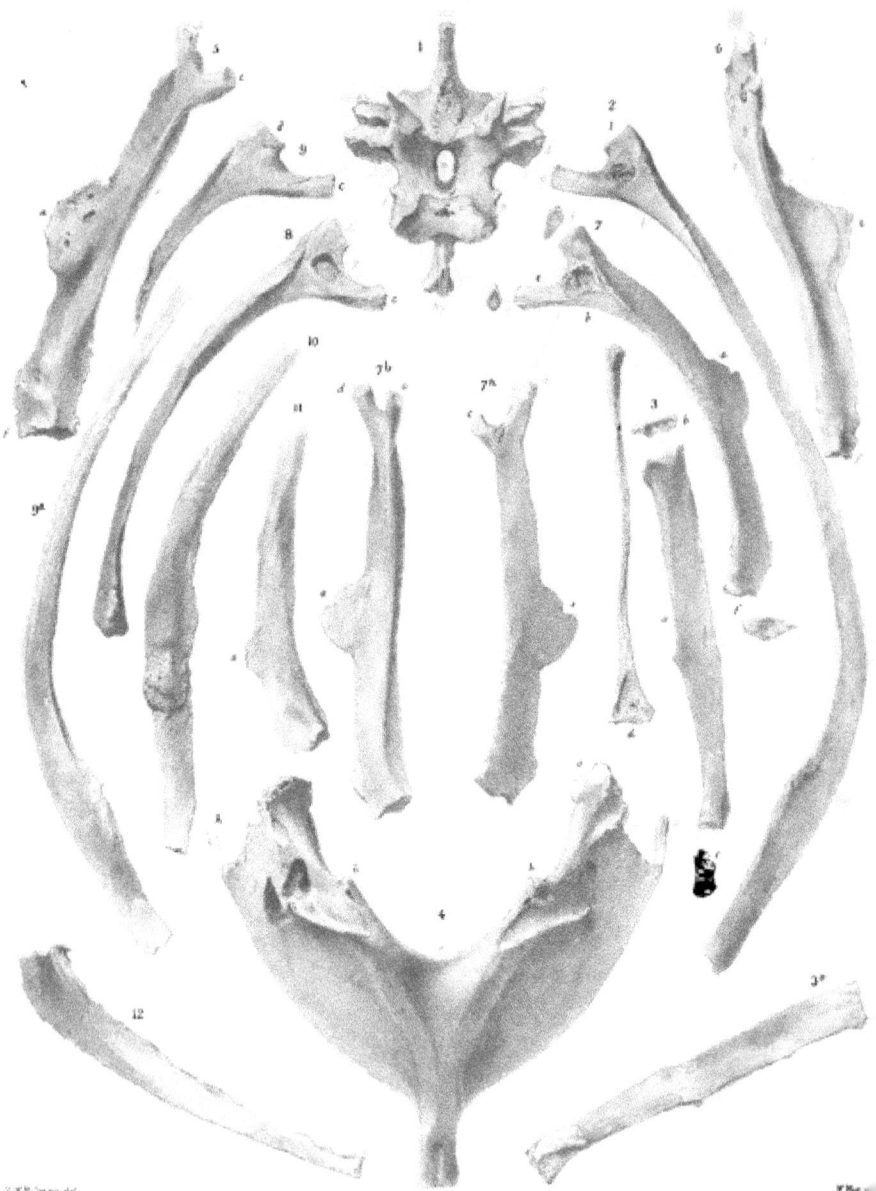

BLANK PAGE

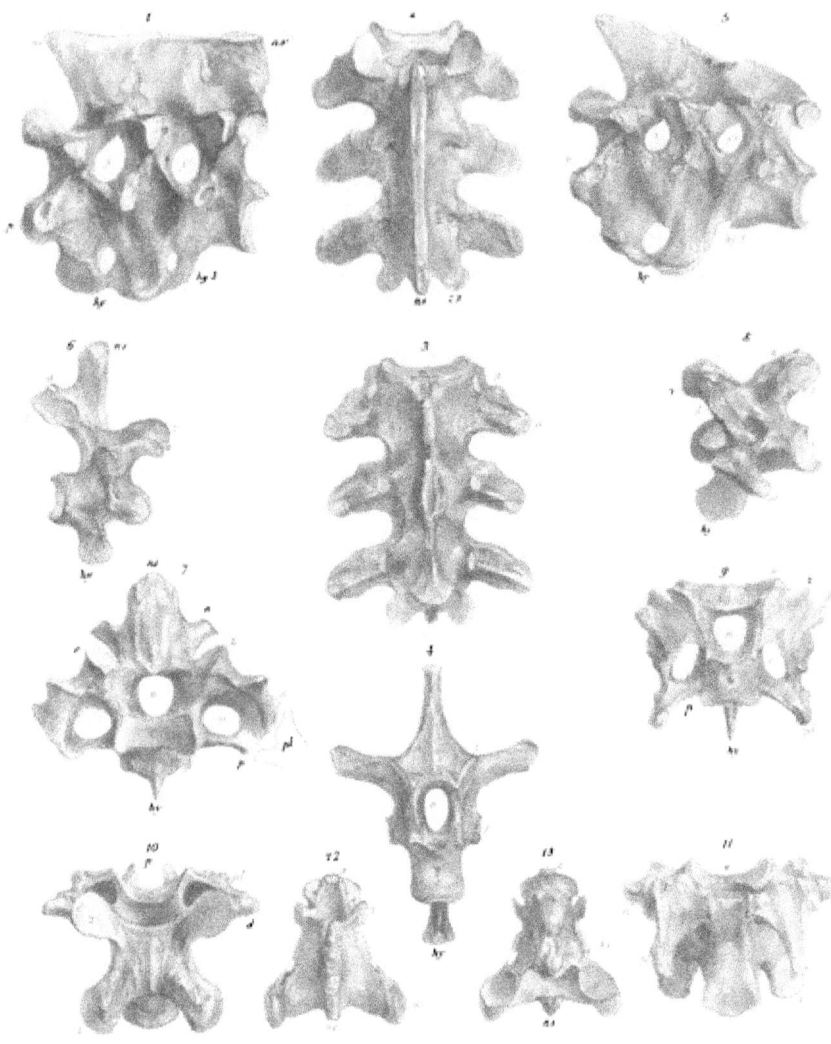

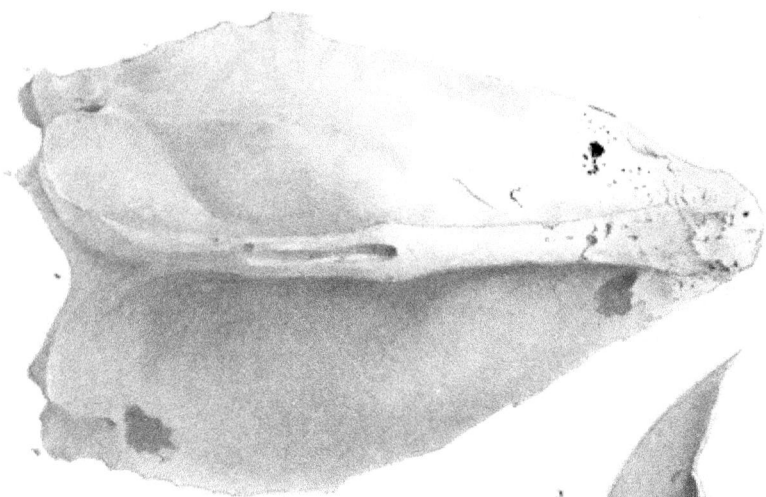

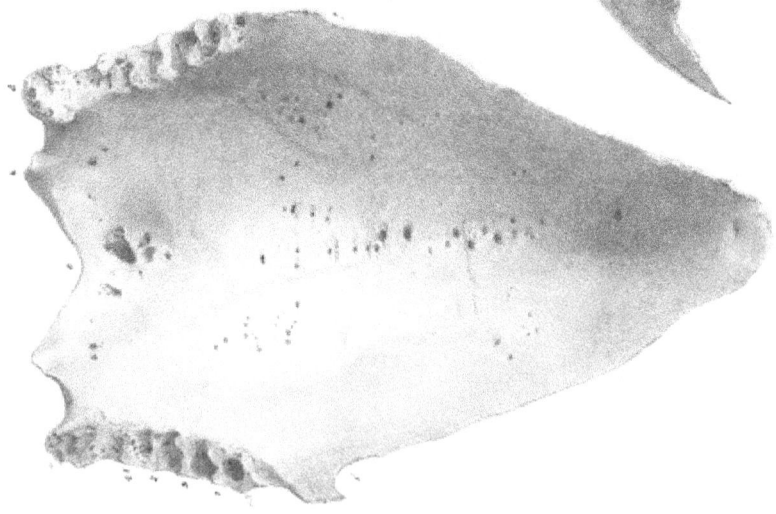

BLANK PAGE

BLANK
PAGE

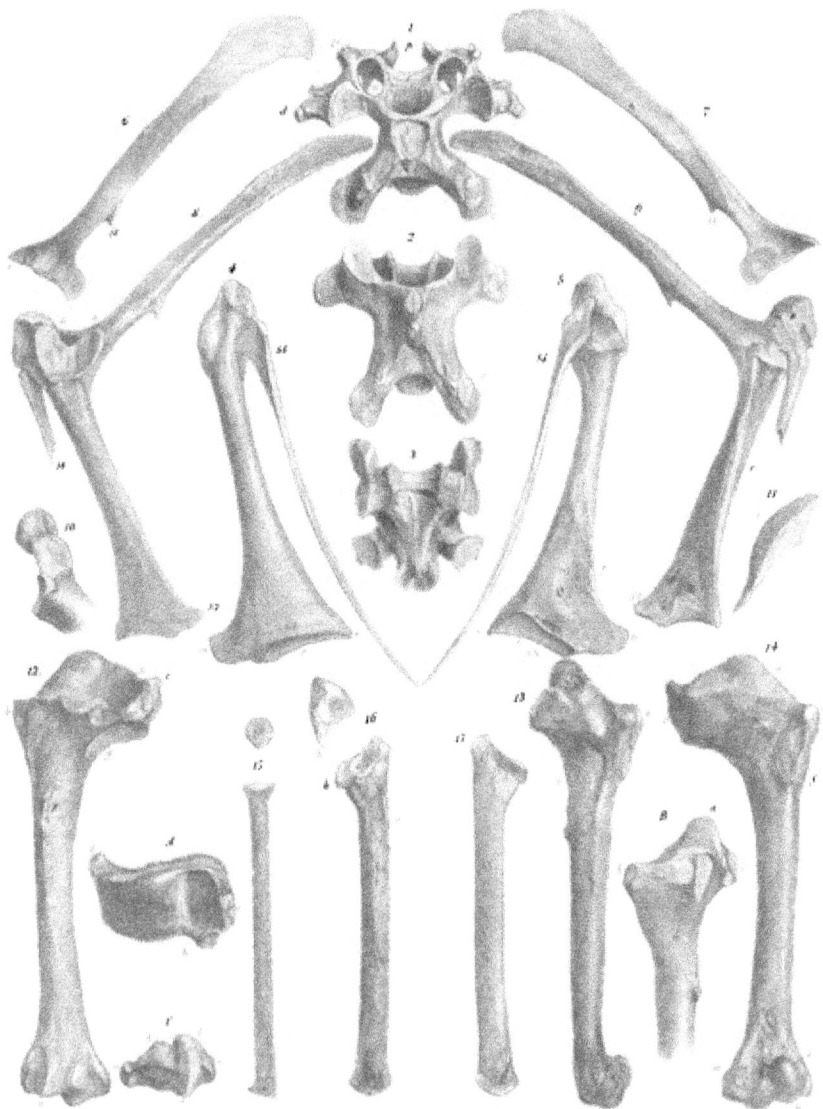

BLANK PAGE

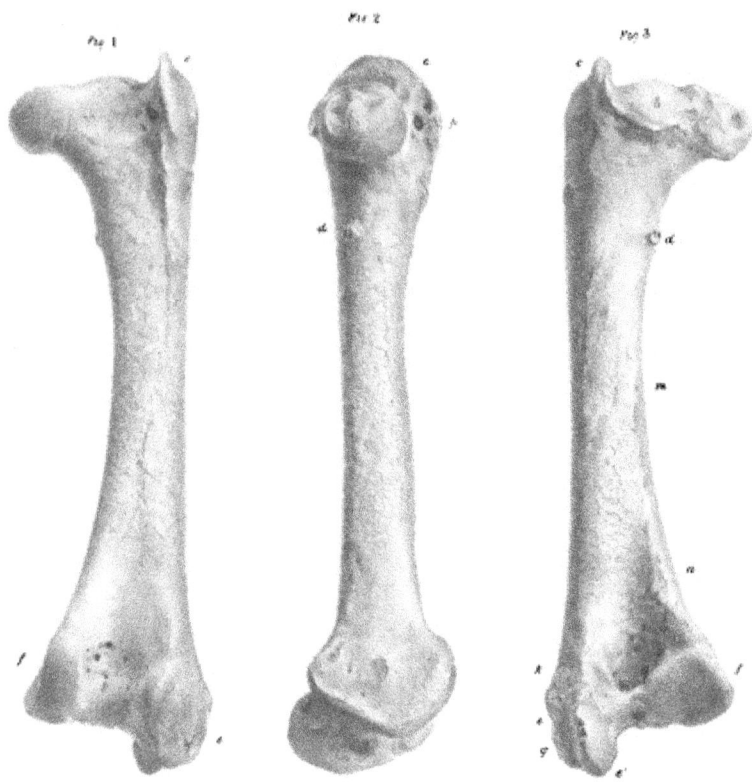

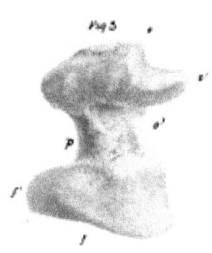

BLANK
PAGE

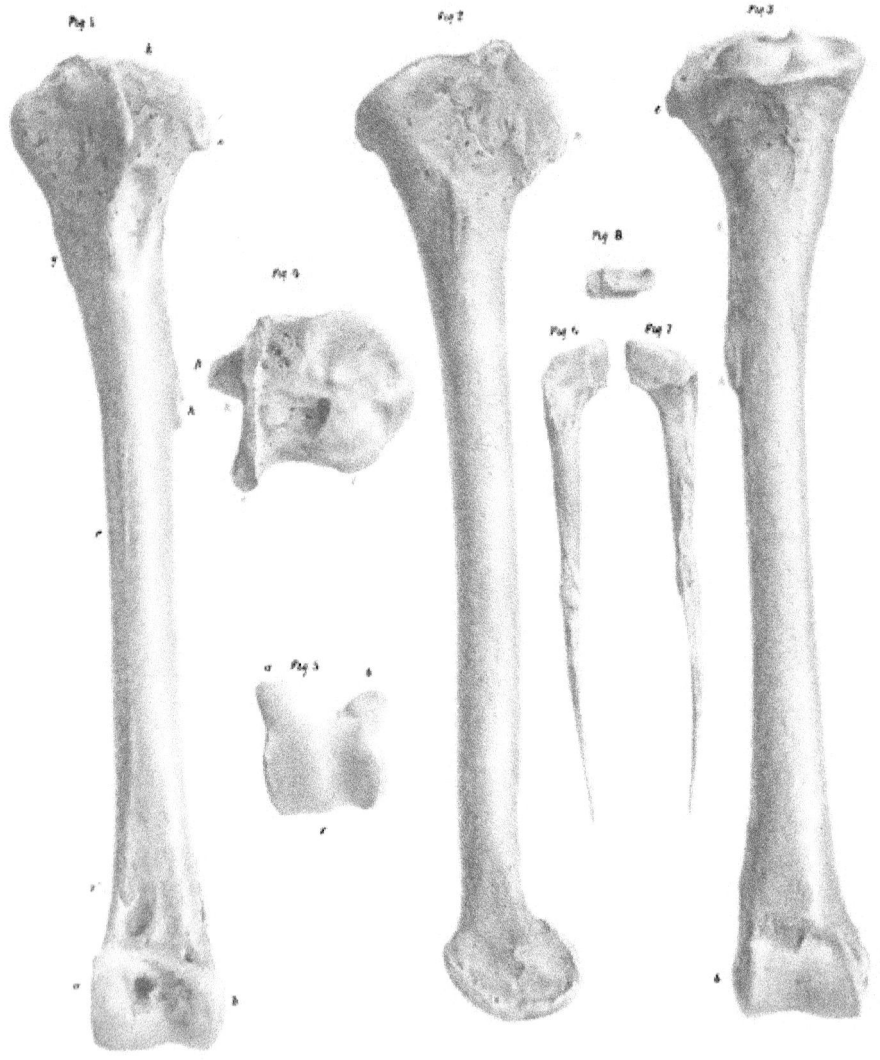

Fig. 1.   Fig. 2.   Fig. 3.

Fig. 8.

Fig. 4.   Fig. 6.   Fig. 7.

Fig. 5.

BLANK
PAGE

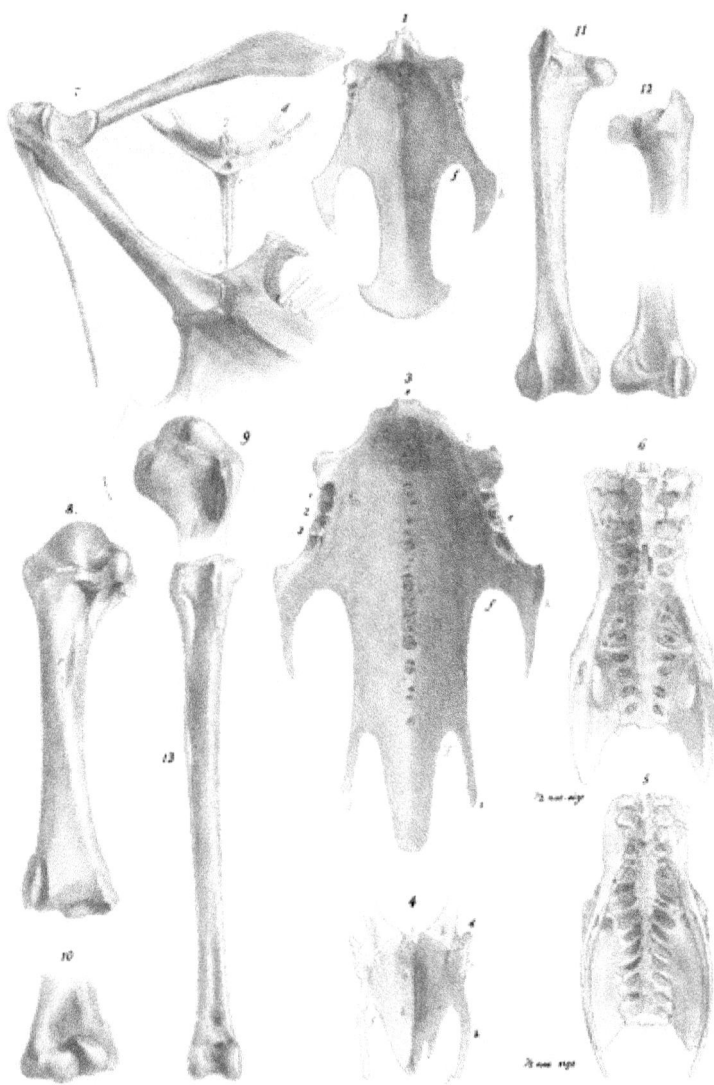

BLANK
PAGE

www.ingramcontent.com/pod-product-compliance
Lightning Source LLC
Chambersburg PA
CBHW022035190326
41519CB00010B/1722